CAMBRIDGE LIBRARY COLLECTION

Books of enduring scholarly value

Philosophy

This series contains both philosophical texts and critical essays about philosophy, concentrating especially on works originally published in the eighteenth and nineteenth centuries. It covers a broad range of topics including ethics, logic, metaphysics, aesthetics, utilitarianism, positivism, scientific method and political thought. It also includes biographies and accounts of the history of philosophy, as well as collections of papers by leading figures. In addition to this series, primary texts by ancient philosophers, and works with particular relevance to philosophy of science, politics or theology, may be found elsewhere in the Cambridge Library Collection.

Naturalism and Agnosticism

James Ward (1843–1925) was Professor of Mental Philosophy and Logic at the University of Cambridge. First published in 1899, this two-volume work consists of his Gifford Lectures, delivered between 1896 and 1898, in which he criticises Naturalism (the belief that all phenomena are governed by the laws of science, and that the supernatural cannot exist), and Agnosticism (the belief that the existence of spiritual phenomena cannot be proved or disproved), in favour of Idealism, in which spiritual and non-material phenomena are central to human experience. The lectures in Volume 1 set Naturalism and Agnosticism within the context of the Mechanical Theory, arguing against its claim that experience can be fully described in terms of mechanical concepts such as motion, energy and force. Exploring the ideas of prominent thinkers such as Newton, Huxley and Spencer, these thought-provoking discussions continue to inform and evoke debate among philosophers and natural scientists.

Naturalism and Agnosticism

The Gifford Lectures Delivered Before the University of Aberdeen in the Years 1896–1898

VOLUME 1

JAMES WARD

CAMBRIDGE UNIVERSITY PRESS

Cambridge, New York, Melbourne, Madrid, Cape Town,
Singapore, São Paolo, Delhi, Tokyo, Mexico City

Published in the United States of America by Cambridge University Press, New York

www.cambridge.org
Information on this title: www.cambridge.org/9781108040976

This edition first published 1899
This digitally printed version 2012

ISBN 978-1-108-04097-6 Paperback

NATURALISM AND AGNOSTICISM

NATURALISM

AND

AGNOSTICISM

THE GIFFORD LECTURES DELIVERED BEFORE
THE UNIVERSITY OF ABERDEEN
IN THE YEARS 1896–1898

BY

JAMES WARD, Sc.D.
HON. LL.D. EDINBURGH
PROFESSOR OF MENTAL PHILOSOPHY AND LOGIC
IN THE UNIVERSITY OF CAMBRIDGE

VOLUME I

"Wer die Gesetzmässigkeit der Natur für das verantwortlich macht, was wirklich geschieht, behauptet damit, dass sie Gedanken realisiere, und ist Teleolog ohne es zu wissen." — SIGWART.

LONDON
ADAM AND CHARLES BLACK
1899

PREFACE

THESE lectures do not form a systematic treatise. They only attempt to discuss in a popular way certain assumptions of 'modern science' which have led to a widespread, but more or less tacit, rejection of idealistic views of the world. These assumptions are, of course, no part of the general body of the natural sciences, but rather prepossessions that, after gradually taking shape in the minds of many absorbed in scientific studies, have entered into the current thought of our time. Though, as I believe, these prepossessions will prove to be ill-grounded and mistaken, yet they are nevertheless the almost inevitable outcome of the standpoint and the premisses from which the natural sciences start. If with the history of science and the results of science before us we pass straight on to the construction of a philosophy, idealism has no chance. But, in truth, 'modern science' hardly needs to *construct* its philosophy; for, without any conscious labour on its part, the naturalistic view of the world seems to stand out clearly of itself. Figuratively speaking we have, as it were, the nebular hypothesis exemplified in the evolution of knowledge. (And for Mr. Spencer, by the way, the exemplification is more than figurative.) From an inchoate confusion of *Glaube* and *Aberglaube*, of probable opinions and fanciful surmises, there gradually emerges the clear circle of the sciences, waxing brighter as it advances in coherence and continuity, while the void of nescience beyond grows too dark for shadows, too empty for dreams; till at length all there is to know finds a place in an unbroken concatenation of laws, binding nature fast in fate. Taking science as the touchstone of knowledge, "knowing in the strict sense," as Mr. Spencer calls it, we must admit that we do not know God or even see room for God at all. Such is the naturalistic contrast of science and nescience, on the strength of which Naturalism takes Agnosticism for an ally. But the agnostic opposition of knowable and unknowable is by no means identical with this contrast; and the alliance is proving ill-starred in consequence. For the distinction of known and unknown, as science intends it, is, we may say, a mere objective distinction of fact; the distinction of knowable and unknowable as used by the agnostic, on the other hand, brings the knower himself to the fore, and entails an examination both of the standpoint and of the premisses from which science, without any preliminary criticism, set forth. In other words, Naturalism is essentially dogmatic, whereas Agnosticism is essentially sceptical.

But this strange *liaison*, though disastrous to Naturalism, has served to promote Idealism in sundry ways. The old materialism has been repudiated and an agnostic or neutral monism — nihilism some would call it — has come into vogue in its stead. 'Modern Science' seems at this point in a dilemma; either this nondescript monism must lapse back into materialism or move on to spiritualism. But the relapse is difficult and the present position unstable. With these more strictly epistemological topics I have tried to deal in the second and shorter half of this work (beginning, that is, with *Lecture XIV*). Many who chance to glance at its contents, especially if they should be students of philosophy, may think that here was the place to begin, and that the earlier and longer division of the book could be suppressed without much detriment to the justification of idealism that follows. That the one half might have been expanded and the other contracted with advantage, I fully admit; and had it been any way practicable to recast lectures, delivered on five separate occasions, into one whole, such a readjustment might have been effected. But, in any case, it would have seemed essential to the writer's argument and purpose to discuss what have been called the real principles of Naturalism at some length.

I take it for granted that till an idealistic (*i.e.* spiritualistic) view of the world can be sustained, any exposition of theism is but wasted labour. Such, at any rate, is the opinion of those who are dominated by naturalistic preconceptions, and that — so far as these discussions are concerned — is sufficient. But now, as already said, it is precisely 'the solid ground of nature' science seems to present that makes idealism appear to the naturalist so fatuous or so futile, 'containing nothing but sophistry and illusion, leading to nothing but obscurity and confusion of ideas.' But is it verily positive, fully orbed reality that modern science sets before us? This is the question that leads us to examine the mechanical theory, the theory of evolution and the theory of psychical epiphenomena, the principles on which this supposed unity and completeness seem mainly to depend. Naturalism, we find, though rejecting materialism, abandons neither the materialistic standpoint nor the materialistic endeavour to colligate the facts of life, mind, and history with a mechanical scheme. But the compact of Naturalism with Agnosticism, like the legendary compacts with the devil, to which Lange happily compares it, costs Naturalism, as it turns out, its entire philosophical existence. In order to be free of 'metaphysical quagmires' such as the ideas of substance and cause, it is led to reject the reality not only of mind, but even of matter; and in this state of ideophobia must collapse, for lack of the very ideas it dreads.

The following is a brief outline of the argument:—
A. i. Mechanics, as a branch of mathematics dealing simply with the quantitative aspects of physical phenomena, can dispense entirely with 'real categories'; not so the mechanical theory of Nature, which aspires to resolve the actual world into an actual mechanism. Homœopathic remedies are the best for that disorder; and, in fact, at the present time mathematicians are, of all men of science, the least tainted with it. An inquiry into the character and mutual relations of Abstract Dynamics, Molar Mechanics, and Molecular Mechanics, seems to shew that the modern dream of a mechanical ἀρχὴ is as wild as the Pythagorean of an arithmetical one. (*Lectures II–VI.*)
ii. A powerful, though unintentional refutation of this theory is furnished by Mr. Herbert Spencer's attempt to base a philosophy of evolution on the doctrine of the conservation of energy. When at length Naturalism is forced to take account of the facts of life and mind, we

find the strain on the mechanical theory is more than it will bear. Mr. Spencer has blandly to confess that 'two volumes' of his 'Synthetic Philosophy' are missing, the volumes that should connect inorganic with biological, evolution. (*Lectures VII–IX.*) Turning to the great work of Darwin, we find, on the one hand, no pretence at even conjecturing a mechanical derivation of life;[1] and, on the other, we find teleological factors, implicating mind and incompatible with mere mechanism, regarded as indispensable. (*Lecture X.*) iii. And finally, when confronted with the relation of mind and body, Naturalism is driven, in the endeavour to maintain its mechanical basis inviolable, to broach psychophysical theories in flagrant contradiction not only with sound mechanical principles and sound logic, but with the plain facts of daily experience. To the body as a phenomenal machine corresponds the mind as an epiphenomenal machine, albeit the correspondence cannot be called causal in any physical sense, nor casual in any logical sense. (*Lectures XI–XIII.*)

B. An examination of the 'real principles' of Naturalism thus secures us a specially advantageous position for discussing the epistemological questions on which the justification of idealism depends. iv. The dualism of matter and mind, which has made the connexion of body and soul an enigma for the naturalist, has rendered the converse problem, as to the perception of an external world, equally vexatious to the psychologist. It is obvious that

[1] "It is mere rubbish thinking at present of the origin of life; one might as well think of the origin of matter." — Letter to Hooker, *Darwin's Life*, vol. iii, p. 18.

there is no such dualism in experience itself, with which we must begin; and reflecting upon experience as a whole, we learn how such dualism has arisen: also we see that it is false. (*Lectures XIV–XVII.*) Further, such reflexion shews that the unity of experience cannot be replaced by an unknowable that is no better than a gulf between two disparate series of phenomena and epiphenomena. Once materialism is abandoned and dualism found untenable, a spiritualistic monism remains the one stable position. It is only in terms of mind that we can understand the unity, activity, and regularity that nature presents. In so understanding we see that Nature is Spirit. (*Lectures XVIII–XX.*)

It is to be feared that inconsistencies and misunderstandings may be detected in the course of an argument elaborated piecemeal over a period of three years, and continually interrupted by other work. Some of these I might myself have discovered had it been possible to do more than publish the lectures substantially as they were delivered.

There only remains the pleasant duty of acknowledging the valuable help received from many kind friends. Among these I must mention Professor Poynting, F.R.S., of Mason's College, Birmingham; Dr. E. W. Hobson, F.R.S., of Christ's College, Cambridge; the Hon. B. A. W. Russell, Fellow of Trinity College; and particularly Professor J. S. Mackenzie, of University College, Cardiff, who has aided me with many judicious criticisms in the course of reading the proof sheets.

JAMES WARD.

TRINITY COLLEGE, CAMBRIDGE,
March, 1899.

CONTENTS

OF THE FIRST VOLUME

LECTURE I

INTRODUCTION

PART I. THE MECHANICAL THEORY

LECTURE II

ABSTRACT DYNAMICS

LECTURE III

RELATION OF ABSTRACT DYNAMICS TO ACTUAL PHENOMENA

LECTURE IV

MOLECULAR MECHANICS: ITS INDIRECTNESS

LECTURE V

MOLECULAR MECHANICS: IDEALS OF MATTER

LECTURE VI

THE THEORY OF ENERGY

PART II. THEORY OF EVOLUTION

LECTURE VII

MECHANICAL EVOLUTION

LECTURE VIII

MR. SPENCER'S INTERPRETATION OF EVOLUTION

LECTURE IX

REFLEXIONS ON MR. SPENCER'S THEORY: HIS TREATMENT OF LIFE AND MIND

LECTURE X

BIOLOGICAL EVOLUTION

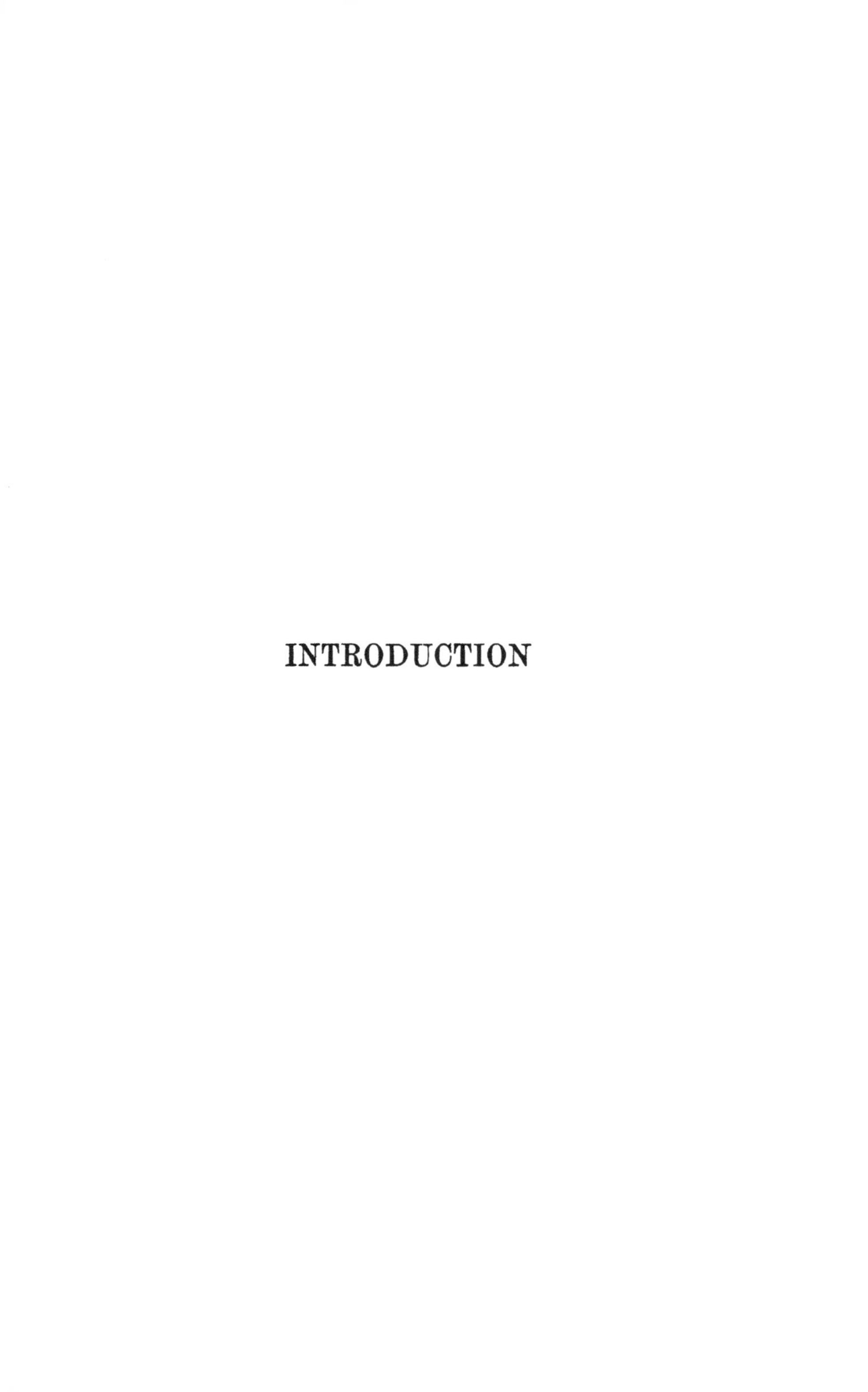

INTRODUCTION

NATURALISM AND AGNOSTICISM

LECTURE I

INTRODUCTION

The attitudes towards Theism of Newton and Laplace: the latter has become the common attitude of 'Science.' This illustrated.

The polity of Modern Science claims to be in idea a complete and compacted whole. 'Gaps,' in what sense admitted, and how dealt with.

The dualism of Matter and Mind: 'Science' decides to treat the former as fundamental, the latter as episodic.

Professor Huxley on the situation: his admissions and advice — a blend of Naturalism and Agnosticism. These doctrines complementary: they react upon each other. According to the one, Natural Theology is unnecessary; according to the other, Rational Theology is impossible.

Examination of the position that Science forms a self-contained whole. No sharp boundary between 'science and nescience.' Mr. Spencer betrays science.

Tyndall's suggestion of an Emotional Theology.

SIR ISAAC NEWTON concludes his famous *Principia* with a general scholium, in which he maintains that "the whole diversity of natural things can have arisen from nothing but the ideas and the will of one necessarily existing being, who is always and everywhere, God Supreme, infinite, omnipotent, omniscient, absolutely perfect." A little more than a hundred years later Laplace began to publish his *Mécanique Céleste*, which may be described as an extension of Newton's *Principia* on

Newton's lines, translated into the language of the differential calculus. When Laplace went to make a formal presentation of his work to Napoleon, the latter remarked: "M. Laplace, they tell me you have written this large book on the system of the universe and have never even mentioned its Creator." Whereupon Laplace drew himself up and answered bluntly: "Sire, I had no need of any such hypothesis."[1] Since that interview another century has almost passed. Sciences that were then in their infancy — such as chemistry, geology, biology, and even psychology — have in the meantime attained imposing proportions. Any one who might now have the curiosity to compare the treatises of their best attested exponents with the great work of Laplace would find that work no longer singular in the omission which Napoleon found so remarkable, an omission which Newton, by the way, in his famous letters to Bentley, had already pronounced to be absurd.

Of course, it is not to be forgotten, the increasing specialisation brought about by the growth of knowledge justifies and even necessitates far greater restriction in the scope of any given branch of it than was customary a couple of centuries ago. People talked then, not of this or that natural science, but of 'natural philosophy'; and psychology, as we know, even in our own day, is often lumped together with metaphysics as 'mental philosophy.' It was incumbent on men styling themselves philosophers to define their attitude towards the notions of a necessarily existent Being, a First and Absolute Cause, and not to confine themselves merely to contin-

[1] W. W. Rouse Ball, *Short History of Mathematics*, 1888, p. 388.

gent existences and to causes that are in turn conditioned. The sharp division which Christian Wolff brought into vogue between empirical and rational knowledge was then ignored, if not unknown. But nowadays, at all events, the absence from a work on natural science of all reference to the supernatural would be no proof that the author disavowed the supernatural altogether.

Still, this is not the point. What we have to note is the existence in our time of a vast circle of empirical knowledge in the whole range of which the idea of a Necessary Being or a First Cause has no place. Towards this result religious and devout men like Cuvier or Faraday have contributed as much as atheists such as Holbach or Laplace. Like many another result of collective human effort, it was neither intended nor foreseen. But there it is nevertheless; and it is all the more impressive because it has grown with humanity, and is not the work of a one-sided sect or school. If modern science had a voice and were questioned as to this omission of all reference to a Creator, it would only reply: I am not aware of needing any such hypothesis.

God made the country, they say, and man made the town. Now we may, as Descartes did, compare science to the town. It is town-like in its compactness and formality, in the preëminence of number and measurement, systematic connexion, and constructive plan. And where science ends, they say too, philosophy and faith may begin. But where *is* science to end? All was country once, but meanwhile the town extends and extends, and the country seems to be ever receding before it. Let us recall a few familiar instances by way of illustration.

To Bentley's inquiry, how the movements and structure of the solar system were to be accounted for, Newton replied: "To your query I answer that the motions which the planets now have could not spring from any natural cause alone, but were impressed by an intelligent Agent. . . . To make this system with all its motions required a cause which understood and compared together the quantities of matter in the several bodies of the sun and planets and the gravitating powers resulting from thence, . . . and to compare and adjust all these things together in so great a variety of bodies, argues that cause to be not blind and fortuitous, but very well skilled in mechanism and geometry."[1] But now, in place of this direct intervention of an intelligent Agent, modern astronomy substitutes the nebular hypothesis of Kant and Laplace. Think again of the remarkable instances of special contrivance and design collected by Paley in his *Natural Theology*, published at the beginning of this century, or of those of the *Bridgewater Treatises* a generation later — works from which some of us perhaps got our first knowledge of science. Nobody reads these books now, and nobody writes others like them. Such arguments have ceased to be edifying, or even safe, since they cut both ways, as the formidable array of facts capable of an equally cogent dysteleological application sufficiently shews. But, in truth, special adaptations have ceased to lie on the confines of science, where natural causes end. "Sturmius," says Paley, "held that the examination of the eye was a cure for atheism."[2] Yet Helm-

[1] Bentley's *Works*, Dyce's edition, vol. iii, pp. 204-206.

[2] *Natural Theology*, ch. iii, Tegge's edition of the *Works*, p. 263.

holtz, who knew incomparably more about the eye than half a dozen Sturms, describes it as an instrument that a scientific optician would be ashamed to make: and Helmholtz was no atheist.[1] Again the immutability and separate creation of species, which Cuvier and other distinguished naturalists long stoutly maintained, are doctrines now no longer defensible. And without them the unique position assigned to man in the scale of organic life — for the sake of which, it is not too much to say, Cuvier and his allies held out so desperately — can be claimed for man no more. "The grounds upon which this conclusion rests," says Darwin, the conclusion, *i.e.*, that man is descended from some less highly organised form, "will never be shaken, for the close similarity between man and the lower animals in embryonic development as well as in innumerable points of structure and constitution, both of high and of the most trifling importance, — the rudiments which he retains, and the abnormal reversions to which he is occasionally liable, — are facts which cannot be disputed."[2] And certainly the unanimity with which this conclusion is now accepted by biologists of every school seems to justify Darwin's confidence a quarter of a century ago. And not merely man's erect gait and noble bearing, but his speech, his reason, and his conscience too, are now held to have been originated in the course of a vast process of evolution, instead of being ascribed, as formerly, to the inspiration and illumination of the Divine Spirit directly intervening.

[1] *Popular Lectures*, 1893, vol. i, p. 194.

[2] *The Descent of Man*, 1871, vol. ii, p. 385.

But vast as the circuit of modern science is, it is still of course limited. On no side does it begin at the beginning, or reach to the end. In every direction it is possible to leave its outposts behind, and to reach the open country where poets, philosophers, and prophets may expatiate freely. However, we are not for the present concerned with this extra-scientific region — the metempirical as it has been called: what we have to notice is rather the existence of serious gaps *within* the bounds of science itself. But over these vacant plots, these instances of *rus in urbe*, science still advances claims, endeavouring to occupy them by more or less temporary erections, otherwise called working hypotheses. Concerning such gaps more must be said presently. Meanwhile, it may suffice to refer to one or two in passing, as our immediate concern is only to understand the claim of science to include them within its domain, though it can only occupy them provisionally.

There is first the great gap between the inorganic and the organic world. Even if astronomical physics will carry us smoothly from chaotic nebulosity to the order and stability of a solar system, and if again "it does not seem incredible that from . . . low and intermediate forms, both animals and plants may have been developed"; [1] still what of the transition from the lifeless to the living? There is no physical theory of the origin of life. Nothing can better shew the straits to which science is put for one than the reception accorded to Lord Kelvin's forlorn suggestion that possibly life was brought to this planet by a stray meteorite!

[1] *Origin of Species*, sixth edition, p. 425.

But, on the other hand, taking living things as there, science finds nothing in their composition or in their processes physically inexplicable. The old theory of a special vital force, according to which physiological processes were at the most only analogous to — not identical with — physical processes, has for the most part been abandoned as superfluous. Step by step within the last fifty years the identity of the two processes has been so far established, that an eminent physiologist does not hesitate to say "that for the future, the word 'vital' as distinctive of physiological processes might be abandoned altogether."[1] It is allowed that life has never been found to arise save through the mediation of already existing life — in spite of many a long and arduous search. Yet, on the ground that vital phenomena furnish no exceptions to purely physical laws, it is assumed that life at its origin — if it ever did originate — formed no break in the continuity of evolution. This instance may perhaps be taken as a type of the scientific treatment of existing lacunæ in our empirical knowledge. Wherever there are reasons for maintaining that a natural explanation is *possible*, though none is, in fact, forthcoming, there actual discontinuity and the supernatural are held to be excluded.

But this principle is put to a far severer trial when we pass from the physical aspect of life to the psychical. The coarse and shallow materialism that disposed of this difficulty with an epigram, "The brain secretes thought as the liver secretes bile," only served to set the

[1] Professor Burdon Sanderson, *Opening Address* to the Biological Section, British Association, 1889. *Nature*, vol. xl, p. 522.

problem in a clearer light. For it is just the hopelessness of the attempt to resolve thought into a physiological function that is the difficulty. And accordingly, within twenty years after Karl Vogt's flippant utterance, we find the physiologist, Du Bois-Reymond, answering this 'riddle,' not merely with an *Ignoramus*, but with an *Ignorabimus*. Indeed, nowadays there is nothing that science resents more indignantly than the imputation of materialism. For, after all, materialism is a philosophical dogma, as much as idealism. It professes to start from the beginning, which science can never do; and, when it is true to itself, never attempts to do. Modern science is content to ascertain coexistences and successions between facts of mind and facts of body. The relations so determined constitute the newest of the sciences, psychophysiology or psychophysics. From this science we learn that there exist manifold correspondences of the most intimate and exact kind between states and changes of consciousness on the one hand, and states and changes of brain on the other. As respects complexity, intensity, and time-order the concomitance is apparently complete. Mind and brain advance and decline *pari passu;* the stimulants and narcotics that enliven or depress the action of the one tell in like manner upon the other. Local lesions that suspend or destroy, more or less completely, the functions of the centres of sight and speech, for instance, involve an equivalent loss, temporary or permanent, of words and ideas. Yet, notwithstanding this close and undeviating parallelism between conscious states and neural states, it is admitted, as I have already said, that the two cannot be identified. It is possible, no doubt, to regard a brain change as a case of matter and motion, but the attempt to conceive a change of mind in this wise is allowed to be ridiculous.

But though these two sets of facts cannot be identified, as the physical and the physiological may be, yet, since they vary concomitantly, may not causal connexion at all events be safely affirmed of them? Yes, it is said, if that means merely that the connexion is not casual. When, however, the attempt is made to determine an event in either as the cause or the effect of the concomitant event in the other, the difficulties seem insuperable. There is not merely the difficulty that the two are strictly coincident in time, so that all question of sequence is excluded — although this difficulty is one on which stress has been laid. But, in addition, the series of neural events — being physical — is already, so to say, closed and complete within itself, each neural state is held to be wholly the effect of the neural state immediately preceding it, and the entire cause of that directly following. In other words, the master generalisation of the physical world, that of the conservation of energy, would be violated by the assumption that energy could appear or disappear in one form without at once disappearing or reappearing to a precisely equivalent amount in another. Brain changes could not then be transformed into sensations, or volitions be transformed into brain changes, without a breach of physical continuity; and of such a breach there is supposed to be no evidence.

The position, then, of science in the present day as regards what I have called the gap between the psychical and the physical is briefly this: If the mechanical theory of the material world including the modern principle of energy is not to be impugned, then there is no natural explanation of the parallelism that exists between processes in brain and processes in consciousness; the gap is one across which no causal links can be traced. This amount of dualism science seems content to admit rather than forego the strict continuity and necessary concatenation of the physical world. But it is not regarded as the sort of discontinuity that sets empirical generalisation at defiance or points directly to supernatural interference. True, the gulf is such that the utmost advance on the physical side would not, of itself, help on psychology in the least, nay would not even suggest to the physicist, pure and simple, the existence of the psychical side at all. True, again, the gap is such that psychology, keeping strictly to its own domain, gives no hint of the existence of that physical mechanism of brain, nerve, and muscle, by which it is so intimately shadowed, or — as many very arbitrarily prefer to say — which it so intimately shadows. But this very concomitance is itself a uniformity of nature, a uniformity of coexistence, and no limit can be assigned to the extent to which psychophysics may succeed in determining its details. Inasmuch as supernatural intervention is not invoked by physiology or psychology severally, psychophysics can obviously dispense with it in merely correlating the two. As a result of our brief survey, then, we find that "the ideas and the will of the one necessarily existing Being," to which Newton referred, do not figure even as a working hypothesis anywhere within the range of that systematic exposition of "the whole diversity of natural things," that calls itself modern science.

This summary of existing knowledge about whatever comes to be is confessedly meagre in the extreme. To many it will suggest objections and to some it may seem obscure. I shall myself have objections in plenty to make and to meet, as best I may, later on; as to the obscurity, this I fear could only be remedied by an elaboration of detail which would call for more time than we can spare. Moreover, this defect is made good already in sundry well-known essays and addresses by men like Huxley, Tyndall, Clifford, Helmholtz, Du Bois-Reymond, and others. Besides, it is precisely the broadest and most general characteristics, not the details, of the current science of nature, that I wish to emphasise. Let me then, before attempting to advance further, ask your patience while I try to restate them in another way.

We note first of all the old dualism of Matter and Mind, or rather — since a duality of substances is nowadays neither asserted nor denied — the dualism of so-called material and mental phenomena. As to material phenomena — that is to say wherever there is matter in motion, whether planets revolving round a sun or molecules vibrating in a living frame, over all these — certain mechanical laws are held to be supreme; that a single atom should deviate from its predetermined course were as much a miracle as if Jupiter should break away from its orbit and set the whole solar system in commotion. Matter and energy are the two fundamental conceptions here. The amount of both is constant, and even independent, in so far as matter cannot be raised to the dignity of energy nor energy degraded to the inertness of matter. But the energy of any given body or material system may vary indefinitely, provided only every increase or decrease shall entail always an equivalent decrease or increase by transfer to or from other bodies or systems. Thus the continuity and solidarity of the material world is complete ; but there is no limit to the diversities which it may assume, provided its physical unity and concatenation are strictly conserved.

When we turn to what are called mental phenomena we find nothing answering to this quantitative constancy, inviolable continuity, and strict reciprocity. Minds are not a single conservative system as matter and energy are. What one mind gains in ideas, feeling, strength of will, another does not necessarily lose. We have here a number of separate individuals, not a single continuum. But on the other hand we know nothing of minds without a living body and without external environment. Between each living body and this environment there is a continuous exchange of material — the metabolism of physiologists — accompanied by a constant give and take of energy. While the organism gains in this exchange, it thrives and developes, goes up in the world ; as it loses, it begins to decline and perish, to go down in the world. But, as all organisms collectively, together with their environments, constitute the constant and continuous physical system, indefinite increase and advance all round are impossible. Sooner or later what we describe as struggle must ensue, leading to 'the survival of the fittest,' as its result. But conscious life is found to rise and fall with organic efficiency and position, so that (completely isolated and distinct as the consciousness of A is from that of B), all minds are indirectly connected; each is yoked to its own body and through this body to the one material world. Of other connexions and relations that minds may have wholly independent of this physical connexion, we have so far no experience; all intercourse, all tradition, is mediated through the one physical world.

So then the concomitance of mind with body is invariable; concomitance of body with mind on the other hand is not certainly more than occasional, even exceptional. Moreover, keeping strictly to the psychological standpoint, we can never get beyond qualitative description and rough classification, natural history in a word, not natural science. And this would be true even though, in *individual* cases, quantitative determinations were possible, which however they are not. For there are certainly no *common* psychological units of intensity or duration; no mind-stuff fixed in amount; no psychical energy that must be conserved. Thus, on the physical side we have a single system, unvarying law, quantitative exactness, complete concatenation of events — in a word, one vastly complex, but rigidly adjusted, mechanism. But on the psychical side we have as many worlds as there are minds, connected indeed, yet independent to an indefinite extent; a series of partial and more or less disparate *aperçus* or outlooks; each for itself a centre of experience, but all without any exact orientation in common. Psychology, pure and simple, has always been individualistic and accepted, tacitly at least, the *Homo Mensura* doctrine. Again, on the physical side the elements with which we deal are held to be indestructible and unalterable, the same always and everywhere. Whereas minds, so far as we know them, are the subjects of continual flux while they last; and seem to arise and melt away like streaks of morning cloud on the stable firmament of blue. But though all these unique and transient centres of thought and feeling are psychologically as isolated and individual as mountain summits, oases in a desert, or stars in space, yet they are indirectly related through physical organisms, which are integral parts of the one great mechanism. To set out, then, from this one permanent material scheme and to trace its working through the fleeting multitude of vital sparks, as one follows the stem of a tree up into its branches with their changing leaves and fruit — that is a sure, synthetic, and direct method. But to attempt, setting out from these sporadic and shifting complexities, to reach an abiding and fundamental unity, is as precarious as analytic and inverse methods always are; and possibly it is altogether hopeless. In brief, then, we are to conclude that, in proportion as psychological facts are physiologically interpretable, and in proportion again as their physiological concomitants are physically explicable, in that same proportion will every fact of mind have a definite and assignable place as an epiphenomenon or concomitant of a definite and assignable physical fact, and our empirical knowledge approximate towards a rounded and complete whole.

No doubt such consummation of natural science is

indefinitely far off. But it is an ideal. Let me cite a single and very eminent witness. "Any one who is acquainted with the history of science," says Professor Huxley, "will admit, that its progress has, in all ages, meant, and now more than ever means, the extension of the province of what we call matter and causation, and the concomitant gradual banishment from all regions of human thought of what we call spirit and spontaneity. . . . And as surely as every future grows out of past and present, so will the physiology of the future gradually extend the realm of matter and law until it is coextensive with knowledge, with feeling, and with action. The consciousness of this great truth," Mr. Huxley believes, "weighs like a nightmare upon many of the best minds of these days. They watch what they conceive to be the progress of materialism in such fear and powerless anger as a savage feels, when, during an eclipse, the great shadow creeps over the face of the sun. The advancing tide of matter threatens to drown their souls; the tightening grasp of law impedes their freedom."[1]

The alarm and perplexity are, in Professor's Huxley's opinion, alike needless; the "strong and subtle intellect" of David Hume, if only we would ponder his words and accept his "most wise advice" would, he thinks, soon allay our fears and give us heart again. The advice is well-known, but as it will fitly introduce a new trait in the modern scientific attitude, the main features of which it is our present business to characterise, I will ask leave to re-quote it. It was in the

[1] *Collected Essays*, Eversley edition, vol. i, pp. 159 ff.

Inquiry concerning the Human Understanding that Hume wrote: "If we take in hand any volume of divinity, or school metaphysics, for instance, let us ask, Does it contain any abstract reasoning concerning quantity or number? No. Does it contain any experimental reasoning concerning matter of fact and existence? No. Commit it then to the flames; for it can contain nothing but sophistry and illusion." How this advice is to dispel perplexity at "the advancing tide of matter and the tightening grasp of law," and how it is to reassure those who are alarmed lest man's moral nature should be debased by the increase of his knowledge, are perhaps not straightway obvious! Well, the comfort consists simply in saying: *After all the knowledge is very superficial and must always remain so.* As Professor Huxley puts it: "What, after all, do we know of this terrible 'matter' except as a name for the unknown and hypothetical cause of states of our own consciousness? And what do we know of that 'spirit' over whose threatened extinction by matter a great lamentation is arising, . . . except that it is also a name for an unknown and hypothetical cause, or condition, of states of consciousness? And what is the dire necessity and 'iron' law under which men groan? Truly, most gratuitously invented bugbears. . . . Fact I know, and Law I know; but what is this necessity save an empty shadow of my own mind's throwing — something illegitimately thrust into the perfectly legitimate conception of law?" "The fundamental doctrines of materialism," continues Professor Huxley, "like those of spiritualism and most other 'isms' lie outside the limits of philosophical inquiry; and David Hume's great service to humanity is his irrefragable demonstration of what these limits are."

In this deliverance of Professor Huxley we have a fragment of that particular 'ism' for which he is proud to be sponsor and which he has christened Agnosticism. It is in fact that doctrine that has led modern science, as I have already remarked, to separate itself from the pronounced materialism and atheism so common in scientific circles half a century or so ago. But it is only in its bearing on the ideal of knowledge just described that agnosticism concerns us at present. Professor Huxley — in this point following the lead of Mr. Herbert Spencer — concludes the consolatory reflections he derives from Hume and returns to his first position in this wise: "It is in itself of little moment whether we express the phenomena of matter in terms of spirit, or the phenomena of spirit in terms of matter — each statement has a certain relative truth. But with a view to the progress of science, the materialistic terminology is in every way to be preferred. For it connects thought with the other phenomena of the universe, . . . whereas, the alternative, or spiritualistic, terminology is utterly barren, and leads to nothing but obscurity and confusion of ideas. Thus there can be little doubt, that the further science advances, the more extensively and consistently will all the phenomena of Nature be represented by materialistic formulæ and symbols."

This 'nightmare' theory of knowledge, as regards its exclusion of everything supernatural or spiritual, thus closely resembles the doctrines which in the seventeenth century they called Naturalism. And the name has recently been revived. But it is important to bear in mind the difference already noted. Naturalism in the old time tended dogmatically to deny the existence of things divine or spiritual, and dogmatically to assert that matter was the one absolute reality. But Naturalism and Agnosticism now go together; they are the complementary halves of the dominant philosophy of our scientific teachers. So far as knowledge extends all is law, and law ultimately and most clearly to be formulated in terms of matter and motion. Knowledge, it is now said, can never transcend the phenomenal; concerning 'unknown and hypothetical' existences beyond and beneath the phenomenal, whether called Matter or Mind or God, science will not dogmatise either by affirming or denying. This problematic admission of undiscovered country beyond the polity of science has tended powerfully to promote the consolidation of that polity itself. Release from the obligation to include ultimate questions has made it easier, alike on the score of sentiment and of method, to deal in a thoroughly regimental fashion with such definite coexistences, successions, resemblances, and differences as fall within the range of actual experience. The eternities safely left aside, the relativities become at once amenable to system. All this is apparent in the passages just quoted from Professor Huxley.

But I pass now to a new point. Agnosticism, we have just seen, has reacted upon naturalism, inducing in it a more uncompromising application of scientific method to all the phenomena of experience. And it will be found that naturalism in its turn has reacted upon agnosticism, inducing in that a more pronounced scepticism, or even the renunciation of higher knowledge as a duty, in place of the bare confession of ignorance as a fact. The contrast between the certainty of science, with its powers of prediction and measurement, and the uncertainty of philosophic speculation, forever changing, but never seeming to advance, has been one source of this agnostic despondency. The long record of attempts that can only appear as failures, the many highly gifted minds, as it seems, uselessly sacrificed in the forlorn enterprise of seeking beneath the veil of things for the very heart of truth — this, when contrasted with the steady growth of scientific knowledge, might well, as Kant puts it, "bring philosophy, once the queen of all the sciences, into contempt, and leave her, like Hecuba forsaken and rejected, bewailing: *modo maxima rerum, tot generis natisque potens — nunc trahor, exul, inops*."[1] But since Kant's day the position of philosophy has become still more desperate. That agnosticism — for such we might call it — by which he himself supplanted the bold but baseless metaphysics of his rationalistic predecessors, is now in turn scouted as transcendental and surreptitious; is charged, that is, with borrowing from experience the very forms on whose strictly *a priori* character it would rest the possibility of experience. By the advance of what has been called metageometry, still more by the advance of experimental and comparative psychology, and by the wide reach of the conception of evolution, science has encroached upon what Kant regarded as the province of the *a priori*. He allowed that all our knowledge begins

[1] *Critique of Pure Reason*, first edition, Pref., p. 3.

with experience and is confined to experience. He allowed that if the several particulars of that experience had been different, as they conceivably might have been, our *a posteriori* generalisations would have varied in like manner. But a spontaneous generation of knowledge from sense particulars without the aid of *a priori* formative processes, was to him as inconceivable as the spontaneous generation of a living object from lifeless matter without the aid of a vital principle. But now that the physical origin of life is regarded as not merely credible but certain, *a priori* forms of knowledge are out of fashion. Kant's position, in a word, is held to be outflanked. There can be no science without self-consciousness; but then this very self-consciousness, it is said, has been evolved by natural processes. Nature herself has polished, and apparently is still polishing, the mirror in which she sees herself reflected. Kant's dialectic against dogmatic metaphysics is thankfully accepted; but his theory of knowledge is held to be superseded by a better psychology and a better anthropology. All this, of course, really amounts to saying that there can be no theory of knowledge at all as distinct from an account of the natural processes by which, as a fact in time, knowledge has come to be. The *solvitur ambulando* procedure is at once the most effective and the most summary method of dealing with this position, and we shall have to try our best at it later on.

Meantime one or two remarks on this unreflective, uncritical, character of modern science may serve to complete this preliminary sketch of its attitude towards the problem of theism. We have seen that, on the one hand, it allows no place for Natural Theology or such knowledge of God as the constitution of nature may furnish; and that, on the other, it denies the title of Knowledge to Rational Theology, or such knowledge of God as philosophy may claim to disclose. We have seen further that these negations have two main grounds: first, the Laplacean *dictum*, which Naturalism adopts, that science has no need of the theistic hypothesis; and secondly, the Humean, or ultra-agnostic, *dictum*, that what is neither abstract reasoning concerning quantity or number, nor yet experimental reasoning concerning matter of fact or existence, can only be sophistry or illusion. Disregarding Hume's somewhat rhetorical phraseology, these two statements amount to saying, first, that there is no knowledge save scientific knowledge, or knowledge of phenomena and of their relations, and secondly, that this knowledge is non-theistic. It is worth our while to note that in a sense both these propositions are true, and *that* is the sense in which science in its every-day work is concerned with them. But again there is a sense in which, taken together, these propositions are not true, but this is a sense that will only present itself to the critic of knowledge reflecting upon knowledge as a whole. Thus it is true that science has no need, and indeed, can make no use, in any particular instance, of the theistic hypothesis. That hypothesis is specially applicable to nothing just because it claims to be equally applicable to everything. Recourse to it as an explanation of any specific problem would involve just that discontinuity which it is the cardinal rule of scientific method to avoid. But, because reference to the Deity will not serve for a physical explanation in physics or a chemical explanation in chemistry, it does not therefore follow that the sum total of scientific knowledge is equally intelligible whether we accept the theistic hypothesis or not. Again, it is true that every item of scientific knowledge is concerned with some definite relation of definite phenomena and with nothing else. But, for all that, the systematic organisation of such items may quite well yield further knowledge which transcends the special relations of definite phenomena. In fact, so surely as science collectively is more than a mere aggregate of items or 'knowledges,' as Bacon would have said, so surely will the whole be more, and yield more, than the mere sum of its parts.

And the strictly philosophical term 'phenomenon,' to which science has taken so kindly, is in itself an explicit avowal of relation to something beyond that is not phenomenal. Mr. Herbert Spencer who, more perhaps than any other writer, is hailed by our men of science as the best exponent of their first principles, is careful to insist upon the existence of this relation of the phenomenal to the extra-phenomenal, noumenal, or ontal. His synthetic philosophy opens with an exposition of this "real Non-relative or Absolute," as he calls it, without which the relative itself becomes contradictory. And when Mr. Spencer speaks of this Absolute as the Unknowable, it is plain that he is using the term 'unknowable' in a very restricted sense. I say this, not merely because he devotes several chapters to its elucidation, for these might have been purely negative; but also because it is an essential part of Mr. Spencer's doctrine to maintain that "our consciousness of the Absolute, indefinite though it is, is positive and not negative";[1] that "the Noumenon everywhere named as the antithesis of the Phenomenon, is throughout necessarily thought of as an actuality";[2] that, "though the Absolute cannot in any manner or degree be known, *in the strict sense of knowing*, yet we find that its positive existence is a necessary datum of consciousness; that so long as consciousness continues, we cannot, for an instant, rid it of this datum; and that thus the belief which this datum constitutes, has a higher warrant than any other whatever."[3] In short the Absolute or Noumenal according to Mr. Spencer, though not known in the strict sense, that is as the phenomenal or relative is known, is so far from being a pure blank or nonentity for knowledge that this phenomenal, which *is* said to be known in the strict sense, is inconceivable without it. It is worth noting, by the way, that 'this actuality behind appearances,' without which appearances are unthinkable, is by Mr. Spencer identified with that 'ultimate verity' on which religion ever insists. His general survey of knowledge then has led this pioneer of modern thought, as he is accounted to be, to reject both the Humean dictum that there is no knowledge save knowledge of phenomena and of their relations, as well as the Laplacean dictum that this knowledge is non-theistic.

But it might be maintained that the several relations

[1] *First Principles*, stereotyped edition, p. 92.

[2] *o.c.*, p. 88. [3] *o.c.*, p. 98.

among phenomena may suffice in their totality to constitute an Absolute. Possibly it may be so; this much remains for the present an open question. But even so, it would still be true that any knowledge of this Absolute would not be phenomenal knowledge. Science, which is chary of all terms with a definitely theistic implication, talks freely of the Universe and of Nature; but I am at a loss to think of any single *scientific* statement that has been, or can be, made concerning either the one or the other. By scientific statement I mean one that having a real import is either self-evident or directly proved from experience.[1]

There is still another possibility, some seem to think, which, however, has not yet been realised, and which indeed, it seems to me, never can be realised. It might conceivably have happened, they say, that our finite knowledge of phenomena proved to be a complete and rounded whole as far as it went, a sort of microcosm within the macrocosm; a model of the whole universe on a scale appropriate to our human faculties, rather than a fragment with hopelessly 'ragged edges.' And spite of the many obstinate questionings that show the contrary, it is far from unusual to find scientific men talking as if this preferable ideal, as some perhaps think it, was the sober fact. Thus Mr. Spencer, though controverting all such views, nevertheless describes "science as a gradually increasing sphere," such that "every addition to its surface does but bring it into wider contact with surrounding nescience." True, this with

[1] Kant's discussion of the cosmological antinomies is instructive here in its method even more than in its results.

Mr. Spencer is only a metaphor, whereas for Comte it was a doctrine; but as metaphor or as doctrine it is widely current and most misleading. Our knowledge is not only bounded by an ocean of ignorance, but intersected and cut up as it were by straits and seas of ignorance; the *orbis scientiarum*, in fact, if we could only map out ignorance as we map out knowledge, would be little better than an archipelago, and would show much more sea than land.

Of course the rejoinder will be made, We admit the intervening streaks and shallows; but here our ignorance, like our knowledge, is only relative, whereas, of the illimitable ocean beyond, our ignorance is absolute and profound. By the help of postulates and generalisations which our perceptive experience confirms, and by the help of hypotheses congruent with our present experiences and verifiable by experiences yet to come, we have completed the circle of the sciences and built up a *Systema Naturæ*. I have endeavoured to describe this system of natural knowledge, as it is commonly conceived by those whose genius and enterprise we have to thank for it. The said fundamental postulates and unrestricted generalisations, the various assumptions consciously or unconsciously made, the hypothetical abstractions by which this unity is secured — to all these we must give our best attention later. For the moment I am concerned only with this one conceit: that the several sciences by their mutual attraction, if I might so say, together form a single whole, *totus teres atque rotundus*, floating in "an interminable air" of pure nescience. But unless we are prepared to repudiate logic altogether, this sharp severance of known and unknown, knowable and unknowable, must be abandoned, so radical are the contradictions that beset it. Where nescience is absolute, nothing can be said; neither that there is more to know nor that there is not. But if science were verily in itself complete, this could only mean that there was no more to know; and then there could be and would be no talk of an environing nescience.

Again, if nescience is real,—is such, I mean, that we are conscious of it,—we must at least know that there is more to know. But how can we know this? To say that we know it because of the incompleteness of the phenomenal relations actually ascertained, may be true enough; but of course such an admission gives up at once the *solid* unity of science as it is and the *utter* vacuity of the opposed nescience. We must suppose then that phenomenal knowledge is regarded as *ideally complete*—the fragments sufficing at least to suggest an outline of the whole, helped out by ultimate generalisations such as the conservations of matter and energy, the principle of evolution, and the like. And if it is still held that there is an endless and impalpable envelope of nescience beyond this ideally perfect sphere of positive knowledge, this can only be because the phenomenal implicates the noumenal; the known and knowable, as Mr. Spencer and others teach, being necessarily related to the 'unknowable,' which means, we must remember, the not strictly knowable. But this doctrine too is fatal to any thoroughgoing dualism of science and nescience; on the contrary, it amounts to a dualism of knowledge. As Mr. Spencer himself says:

"The progress of intelligence has throughout been dual. Though it has not seemed so to those who made it, every step in advance has been a step towards both the natural and the supernatural. The better interpretation of each phenomenon has been, on the one hand, the rejection of a cause that was relatively conceivable in its nature but unknown in the order of its actions, and, on the other hand, the adoption of a cause that was known in the order of its actions but relatively inconceivable in its nature. . . . And so there arise two antithetical states of mind, answering to the opposite sides of that existence about which we think. While our consciousness of Nature under the one aspect constitutes Science, our consciousness of it under the other aspect constitutes Religion."[1]

Finally, if on the other hand, it be held that phenomenal knowledge, when ideally complete, will be clear of these noumenal and supernatural implications, then this position again is incompatible with a dualism between science and nescience. For if the sphere of science were so complete as to be clear of all extra-scientific implications, then, as I have already said, there would be no nescience. If, however, there must be nescience so long as science is finite and relative, then *so long* the metaphysical ideas of the Absolute and the Infinite will transcend the limits of actual science, and yet will have a place within the sphere of science ideally complete. In other words, ideally complete science will become philosophy. This conceit or doctrine of an absolute boundary between science and nescience and the en-

[1] *First Principles*, p. 106 *fin.*

deavour to identify with it a like sharp separation between empirical knowledge and philosophic speculation may then, we conclude, be both dismissed as "sophistical and illusory." Nevertheless, as I have said, these notions are widely current in one shape or other, save among the few in these days, who have even a passman's acquaintance with the rudiments of epistemology. One of the most plausible and not least prevalent forms of this doctrine is embodied in the shallow Comtian 'Law of Development,' according to which there are three stages in human thought, the theological, the metaphysical, and the positive; the metaphysical superseding the theological and being in turn superseded by the positive or scientific. A glance at the past history of knowledge would shew at once the facts that make these views so specious and yet prove them to be false.

And now to resume what has been said, and to conclude: I have tried to present an outline sketch of that polity of many mansions, which we may call the Kingdom of the Sciences, and the mental atmosphere in which its citizens live. As the constant inhabitants of large towns, though familiar with shops supplying bread and beef, know nothing of the herds in the meadows or the waving fields of wheat, so the mere *savant* is familiar with 'phenomena and their laws' and with the methods by which they are severally measured and ascertained, but rarely or never thinks of all that 'phenomena' and 'law' and 'method' imply. As a knowledge of what is thus beyond his purview cannot be attained by experiment or calculation, it should surprise us as little to find him associate it with nescience as it surprises us to find the urchins in a slum confusing with the tales of fairy-land what we may try to tell them of the actual facts of country life.

Indeed the resemblance in the two cases is closer than at first it seems. For it is very common for those who decline to recognise Natural or Rational Theology to speak with fervour of what I think we might fairly call Æsthetic Theology. Tyndall, for example, in his once famous Belfast Address to the British Association, spoke thus to the assembled representatives of science: "You who have escaped from these religions into the high-and-dry light of the intellect may deride them; but in so doing you deride accidents of form merely, and fail to touch the immovable basis of the religious sentiment in the nature of man. To yield this sentiment reasonable satisfaction is the problem of problems at the present hour."[1] It seems clear that in Tyndall's opinion this reasonable satisfaction could not need, at any rate, must not have, an intellectual basis either 'high-and-dry,' or otherwise. For he proceeds to describe this religious sentiment as "a force, mischievous, if permitted to intrude on the region of *knowledge*, over which it holds no command, but capable of being guided to noble issues in the region of *emotion*, which is its proper and elevated sphere." Yet a page or two further on Tyndall brings his address to a close with these words: "The inexorable advance of man's understanding in the path of knowledge, and those unquenchable claims of his moral and emotional nature which the understanding can never satisfy, are here equally set forth. The world

[1] Reprint of Address, 1874, p. 60.

embraces not only a Newton, but a Shakespeare—not only a Boyle, but a Raphael—not only a Kant, but a Beethoven—not only a Darwin, but a Carlyle. Not in each of these, but in all, is human nature whole. They are not opposed, but supplementary; not mutually exclusive, but reconcilable. And if, unsatisfied with them all, the human mind, with the yearning of a pilgrim for his distant home, will still turn to the Mystery from which it has emerged, seeking so to fashion it as to give unity to thought and faith; so long as this is done, not only without intolerance or bigotry of any kind, but with the enlightened recognition that ultimate fixity of conception is here unattainable, and that each succeeding age must be held free to fashion the Mystery in accordance with its own needs—then, casting aside all the restrictions of Materialism, I would affirm this to be a field for the noblest exercise of what, in contrast with the *knowing* faculties, may be called the *creative* faculties of man."

I am really at a loss to know whether this is to be taken for climax or anti-climax, pathos or bathos. But of one thing I am sure: tried by the "high-and-dry light of the intellect" this specimen of Professor Tyndall's "eloquence and scientific fire," as the *Saturday Review* called it, is almost too flimsy for derision.

Surely the late professor must have thought lightly of his own teaching, to be ready under the influence of an emotional yearning to cast aside the doctrine to which an "intellectual necessity" (p. 55) had led him, the doctrine by which he discerned in matter "the promise and potency of all terrestrial life"; nay, further, to be ready to refashion the Mystery from which the human mind has emerged so as to give unity to thought and faith. If religious sentiments must not be permitted to intrude on the region of knowledge, how is the refashioning in the interests of this unity to begin? And if nothing short of *creative* faculties can satisfy this sentiment, what about 'the danger' and 'the mischief' to the work of the *knowing* faculties when such sentiment does intrude?

Professor Tyndall does not tell us where he went for his psychology. But Mr. Spencer, to whom he frequently refers, would have taught him that no sentiments are entirely without a cognitive basis, the religious perhaps least of all. This cognitive element in religious sentiment is of necessity amenable to intellectual challenge, just because it is itself of necessity intellectual. No doubt, "ultimate fixity of conception is here unattainable"; but when Professor Tyndall tells us this, has he forgotten that on the very same page he has also declared "it certain that [scientific] views will undergo modification"? In fact, just as religious sentiment implies knowledge, so too do the high-and-dry constructions of the intellect involve "creative faculties"; finality will be impossible and reconstruction a necessity in both regions so long as we only "know in part." But why do I talk of the regions of knowledge? The semblance of two regions, one pure fact, the other pure fancy, one all science, the other all nescience, is just the error that I have been trying to expose and to which this utterly unscientific notion of an emotional theology is due.

PART I

THE MECHANICAL THEORY

THE MECHANICAL THEORY

LECTURE II

ABSTRACT DYNAMICS

The demurrer of modern scientific thought, though illegitimately, yet practically, forecloses theistic inquiries. A discussion of its fundamental positions therefore called for in the interest of such inquiries.

Natural knowledge to be examined (i) formally *as knowledge*, (ii) *as a body of* real *principles. Beginning with the latter, we have (a) the mechanical theory of Nature, (b) the theory of Evolution, and (c) the psychophysical theory of Body and Mind.*

A. The Mechanical Theory: — *The Laplacean calculator; different views of him; he excludes the teleological. Abstract dynamics, a strictly mechanical science, the basis of this theory, which thus divests itself of the real categories of Substance and Cause, and substitutes for them the quantitative terms 'Mass' and 'Force' (or Mass-acceleration). But if this be so, Laplace's calculator never attains to real knowledge.*

ANY attempt in these days to discuss the problem of theism is, we have seen, liable to demurrers more or less emphatic from what we may fairly call the spirit of the age. Naturalism, speaking in the name of science, declares the problem superfluous, and agnosticism, professing to represent reason, declares it to be insoluble. This attitude we have traced to that positivist conception of knowledge which the rapid advances of science and the repeated failures of philosophy have jointly encouraged. Referring to this conception G. H. Lewes remarks:

"A new era has dawned. For the first time in history an explanation of the world, society, and man is presented which is thoroughly homogeneous and at the same time thoroughly in accordance with accurate knowledge; having the reach of an all-embracing system, it condenses human knowledge into a Doctrine, and coördinates all the methods by which that knowledge has been reached, and will in future be extended. . . . Its basis is science — the positive knowledge we have attained, and may attain, of all phenomena whatever. Its superstructure is the hierarchy of the sciences, *i.e.*, that distribution and coördination of general truths which transforms the scattered and independent sciences into an organic whole, wherein each part depends on all that precede and determines all that succeed."[1] In the last lecture we made a cursory examination of this *soi-disant* organic whole of phenomenal knowledge. Even that brief survey would justify us in saying: First, that it is not in itself a homogeneous and organic whole; for the dualism of matter and mind, at any rate, runs through it, and is only evaded by desperate means. Materialism itself is repudiated, but the materialistic terminology is retained as primary and fundamental. Secondly, that it is not a whole of accurate, positive, knowledge; for it confessedly involves postulates and methods, which it is the business of no one of 'the scattered and independent sciences' to scrutinise, and which they all alike, therefore, accept in a naïve and uncritical fashion. Finally, that it is not an all-embracing system. Hamilton has supplied it with a Virgilian motto: *Rerumque* ignarus, *imagine* gaudet. The 'accu-

[1] *History of Philosophy*, 3d edition, vol. ii, p. 590.

rate and strict' knowledge of appearances implicates an indefinite but still positive consciousness of an ultimate Reality; for, says Mr. Spencer, "it is rigorously impossible to conceive that our knowledge is a knowledge of Appearances only without at the same time conceiving a Reality of which they are appearances, for appearance without reality is unthinkable."[1]

But since the theistic problem deals primarily with spirit, not with matter, since further it involves those fundamental principles of knowledge which science is not concerned to discuss, and since finally it belongs to that extra-scientific or supernatural region of 'nescience' which science allows to be, but to lie forever beyond its pale, we might, if so disposed, reasonably contend that the demurrer both of Naturalism and of Agnosticism is altogether *ultra vires;* we might politely request science to mind its own business and proceed at once to our own. In so doing, too, we could safely count on the approval and good-will of many eminent representatives of science in every department. For, after all, agnosticism and naturalism are not science, but, so to say, a philosophy of knowing and being which is specially plausible to, and hence is widely prevalent among, scientific men. But just for this reason it would ill become us to treat them with cavalier disdain. If Gifford Lectures were less numerous, I might not perhaps be justified in devoting a whole course to these initial objections; but as every university in Scotland has always its Gifford Lecturer, I venture to think such restriction is not only allowable but desirable.

[1] *First Principles*, p. 88.

Our knowledge of nature, as unified and systematised according to the naturalistic scheme, may be considered from two sides. We may examine it *formally*, as knowledge, in respect, that is to say, of its postulates, categories, and methods. Or, taking these for granted, as science itself does, we may examine those of its *real* principles to which its supposed unity and completeness are ascribed. Some odd instances of confusion could be cited due to a mingling of these two points of view — a favourite practice with those who, like Huxley and Tyndall, are at once fervent naturalists and pronounced agnostics. We may know where we are when matter is spoken of throughout as an objective fact, or throughout as a mental symbol, but it is bewildering to find it posing in both characters at once. To begin with, let us then, postpone any attempt to get behind the plain deliverances of science by epistemological reflexions; let us give our attention first to its real principles.

There are three fundamental theories which — as we have already noted — are held to be primarily concerned in the unity of nature: *the mechanical theory*, this comes first and 'determines all that succeed'; *the theory of evolution*, which essays in terms homogeneous with this to 'formulate' the development of the world, society, and man; last, *the theory of psychological parallelism*, dealing with the relation of body and mind. To the first of these we may now pass.

There is a well-known passage at the beginning of Laplace's essay on Probability, which may serve as a basis for the remarks I have to offer on the MECHANICAL THEORY OF NATURE. Having enounced as an axiom —

known, he says, as *the principle of sufficient reason*, that "a thing cannot begin to be without a cause to produce it," and having summarily disposed of the notion of free-will as an easily explained illusion, Laplace proceeds: "We ought then to regard the present state of the universe as the effect of its antecedent state and as the cause of the state that is to follow. An intelligence, who for a given instant should be acquainted with all the forces by which nature is animated and with the several positions of the beings composing it, if further his intellect were vast enough to submit these data to analysis, would include in one and the same formula the movements of the largest bodies in the universe and those of the lightest atom. Nothing would be uncertain for him; the future as well as the past would be present to his eyes." "The human mind," he continues, "in the perfection it has been able to give to astronomy, affords a feeble outline of such an intelligence. Its discoveries in mechanics and in geometry, joined to that of universal gravitation, have brought it within reach of comprehending in the same analytical expressions the past and future states of the system of the world. . . . All its efforts in the search for truth tend to approximate it without limit to the intelligence we have just imagined." So wrote Laplace in 1812, and his words have been classic among men of science ever since. As one instance among many shewing in what sense they have been understood, I may mention the Leipzig Address to the Deutscher Naturforscher Versammlung by Émile du Bois-Reymond, an address that has made more stir in its way than Tyndall's Belfast Address of a year or two later, which it seems to have inspired. "As the astronomer," said the Berlin professor, "has only to assign to the time in the lunar equation a certain negative value to determine whether as Pericles embarked for Epidaurus there was a solar eclipse visible at the Piræus, so the spirit imagined by Laplace could tell us by due discussion of his world-formula who the man with the iron mask was or how the *President* came to be wrecked. As the astronomer foretells the day on which — years after — a comet shall reëmerge in the vault of heaven from the depths of cosmic space, so that spirit would read in his equations the day when the Greek cross shall glance again from the mosque of St. Sophia or England have burnt her last bit of coal. Let him put $t = -\infty$ and there would be unveiled before him the mysterious beginning of all things. Or if he took t positive and increasing without limit, he would learn after what interval Carnot's Law will menace the universe with icy stillness. To such a spirit even the hairs of our heads would all be numbered and without his knowledge not a sparrow would fall to the ground."[1]

Spite of these scriptural allusions, it would be a mistake to imagine any connexion between the knowledge of this Laplacean intelligence and Divine omniscience. How God knows, or even what knowledge means when attributed to the Supreme Being, few of us will pretend to understand. But this imaginary intelligence of Laplace knows, *as we know*, by calculation and inference based on observation. To God the secret thoughts of

[1] *Ueber die Grenzen des Naturerkennens*, 4te Aufl., p. 6.

man's heart are supposed to lie open; from this Laplacean spirit they would be forever hidden, were it not that he can calculate the workings of the brain. Human free will and divine foreknowledge have been held to be not incompatible: but free will and mechanical prediction are avowedly contradictory. Laplace therefore is careful to exclude free will. Before the future can be in this way deduced from the past, all motives must admit of mechanical statement and the motions of matter and its configurations be the sole and sufficient reasons of all change.

It would be a mistake again to confound this mechanical theory of the universe with doctrines such as those of Newton, Clarke, Butler, Chalmers, and other Christian apologists. They too refer to events in the material world as "brought about, not by insulated interpositions of divine power exerted in each particular case, but through the establishment of general laws."[1] But they none the less regard the laws and properties of matter as but "the instruments with which God works."[2] Such language may be open to serious criticism, but that just now is not the point. It is enough if we realise that whoever holds the notion of the Living God as paramount can never maintain that exact acquaintance with his instruments is enough to make plain all that God will do or suffer to be done. Thus Newton, at the close of his *Opticks*, declares that the various portions of the world, organic or inorganic, "can be the effect of nothing else than the wisdom and skill of a powerful

[1] Whewell, *Bridgewater Treatise*, 1847 edition, p. 356.
[2] *o.c.*, p. 357.

ever-living Agent who, being in all places, is more able by his will to move the bodies within his boundless uniform *sensorium*, and thereby to form and re-form the parts of the universe than we are by our will to move the parts of our own bodies." To men like Laplace and the French Encyclopædists, of course, this bold anthropomorphism would mean nothing; such strictly voluntary movement being for them a delusion. But coming from Newton, who did not regard man as a machine or conscious automaton, these words shew plainly that *he* would not have subscribed to the mechanical theory, although he laid what are taken to be its foundations.

I must confess to some surprise on finding Jevons, who must certainly be counted on the theistic side as a strenuous opponent of naturalism, nevertheless seeming to approve of Laplace's "mechanical mythology," as it has been called. "We may safely accept," says Jevons, "as a satisfactory scientific hypothesis the doctrine so grandly put forth by Laplace, who asserted that a perfect knowledge of the universe, as it existed at any given moment, would give a perfect knowledge of what was to happen henceforth and forever after. Scientific inference is impossible, unless we may regard the present as the outcome of what is past, and the cause of what is to come. To the view of perfect intelligence nothing is uncertain."[1] I must again repeat, that neither the intelligence conceived by Laplace, nor the knowledge attributed to it, is in any sense entitled to be called perfect. Laplace himself, though accounted hardly second to Newton as a mathematician, was hopelessly incom-

[1] *Principles of Science*, 2nd edition, p. 738.

petent in the region of moral evidence. After a few weeks in office as Minister of the Interior, his master Napoleon sent him about his business,[1] declaring him fit for nothing but solving problems in the infinitely little. His imaginary intelligence was only an indefinite magnification of himself, commanding an appalling amount of differential detail and possessed of the means of integrating it; but there is nothing to shew that the *in*capacity of this Colossus may not in other respects have been as sublime as his capacity for calculation. Jevons's inconsequence in accepting this Laplacean conceit is possibly due to a misunderstanding. A reference to Newton's first law of motion will make my meaning clear. When it is there said that a body *left to itself* perseveres in its state of rest or of uniform motion in a straight line, what is affirmed is a tendency, not a fact. Similarly it might be said of the material universe, that, if from any given moment it were *left to itself*, its state thenceforth and ever after would be the outcome of its state at such initial moment. So understood, Laplace's 'doctrine' would formulate a tendency, but would not assert a fact. That it is in the former sense that Jevons interprets it is plain, for he says expressly: "The same Power, which created material nature, might, so far as I can see, create additions to it, or annihilate portions which do exist. . . . The indestructibility of matter, and the conservation of energy, are very probable scientific hypotheses, which accord satisfactorily with experiments of scientific men during a few years past; but it would be a gross misconception of scientific inference to

[1] Whewell, *o.c.*, p. 338.

suppose that they are certain in the sense that a proposition in Geometry is certain."[1] But this was assuredly not Laplace's meaning; and from the illustrations used it was clearly not what Du Bois-Reymond understood him to mean. And lastly, it is certainly not in any such tentative and provisional sense that the mechanical theory now holds sway among scientific men and 'weighs,' as Huxley put it, 'like a night-mare' on the minds of many.

We are bound, I think, carefully to distinguish these two views: the one regarding the universe, so far at least as we can know it, as a vast automatic mechanism, and the other regarding the 'laws of nature' as but the instrument of Nature's God. But it is important to observe, too, that they have a certain common ground in the recognition of laws as 'secondary causes.' In this the naturalism of modern science and the supernaturalism of popular theology are so far at one; although the naturalist stops at the laws, and the theologian advances to a Supreme Cause beyond them and distinct from them. Now, it is, I think, inevitable, so far as the question of theism is argued out from such premisses, that theism will get the worst of it. Unquestionably it has had the worst of it on these lines so far; of this we noted many instances in the last lecture. Not a few temples to the Deity founded on some impressive fact supposed to be safely beyond the reach of scientific explanation have been overtaken and secularised by the unexpected extension of natural knowledge. Chalmers's now classic distinction between

[1] Whewell, *o.c.*, p. 766.

the laws and the collocation of matter, familiar at least to every reader of Mill's *Logic*, may serve to illustrate this point. "The tendency of atheistical writers," says Chalmers, "is to reason exclusively on the laws of matter, and to overlook its dispositions. Could all the beauties and benefits of the astronomical system be referred to the single law of gravitation, it would greatly reduce the strength of the argument for a designing cause."[1] "When Professor Robison felt alarmed by the attempted demonstration of Laplace, that the law of gravitation was an essential property of matter, lest the cause of natural theology should be endangered by it, he might have recollected that the main evidence for a Divinity lies, not in the laws of matter, but in the collocations."[2] "Though we conceded to the atheist the eternity of matter and the essentially inherent character of all its laws, we would still point out to him, in the manifold adjustments of matter, its adjustments of place and figure, and magnitude, the most impressive signatures of a Deity."[3] But what would become of this 'main evidence for a Divinity' if the laws of matter themselves explained its collocations? They can never explain them completely, of course. Till a definite configuration is given him the physicist has no problem; but with such data he professes to deduce the motions and redistributions that according to the laws of matter must ensue. So, if science by the help of these laws should trace the course of the universe backwards, it must halt at some

[1] *Bridgewater Treatise*, vol. i, p. 17.
[2] *o.c.*, p. 20, *note*. [3] *o.c.*, p. 21.

configuration or other; and of the configuration at which it halts it can give no account. "The laws of nature," says Chalmers, "may keep up the working of the machinery — but they did not and could not set up the machine."[1] This final configuration reached by the scientific regress, then — let it be noted — is "the machine." That — provisionally at all events — science cannot explain; so much is true. But meanwhile two things are noteworthy. First, in innumerable cases, as I have said, what was formerly taken to be part of the machine turns out to be due to the workings of its machinery. Secondly, as a consequence of this, those constructive interventions, which are held "to demonstrate so powerfully the fiat and finger of a God," become rapidly fewer in number, and recede farther and farther into the deep darkness of the infinite past. It was surely a short-sighted procedure to rest the theistic argument on a view of nature that must inevitably reduce the strength and diminish the impressiveness of that argument at every advance of natural science. When, too, those who adopt such a line of reasoning themselves allow this fatal weakness, as we have seen that Chalmers did, the proceeding becomes almost fatuous. Indeed, it would hardly be an exaggeration to say that the naturalism of to-day is the logical outcome of the natural theology of a century ago. I do not forget a rejoinder on the old lines that one frequently hears now that the theories of Lyall and Darwin and Spencer are supposed to have become established truths — a sort of *dernier ressort* where

[1] *Bridgewater Treatise*, vol. i, p. 27.

direct attacks have failed. After all, it is said, the more a machine can direct itself and repair itself the more wonderful its first construction must have been. To have so created and disposed the primal elements of the world as to insure by the steadfast working of unvarying laws the emergence in due time of all the life and glory of the round ocean and the teeming earth, is not this after all "the most impressive signature of a Deity"? This seems to me very like asking whether, after all, infinity times nothing is not greater than *n* times *m*? In other words such an argument points logically either to the machine being nothing and God all, or to God being nothing and the machine everything. But which? That depends where we start: if from God, the machine is throughout dependent; but if from the machinery, we may never reach God at all. For the avowed pantheist, who knows neither secondary laws nor machinery, it is, of course, God that is all.

> "That God, which ever lives and loves—
> One God, one law, one element,
> And one far-off divine event."

For those, on the other hand, anxious, perhaps, like Chalmers, to keep clear of what he calls 'the metaphysical obscurity' concerning the origination of matter and its essential properties, and content to "discern in the mere arrangements of matter the most obvious and decisive signatures of the artist hand which has been employed in it,"[1] for such, it is God that vanishes.

[1] *Bridgewater Treatise*, vol. i, p. 25.

Logically and actually on their premisses we find in the words of Huxley already quoted "that matter and law have banished spirit and spontaneity."[1]

This then is the Laplacean conception that we have first to examine, and if it lead us in the end into 'metaphysical obscurity,' let us be warned not to shrink from the task. In the beginning, however, it will rather be certain physical commonplaces that must claim our attention. As to these it behoves me to say at once and emphatically that I make no pretence to special knowledge. But I shall take care to refer to nothing — unless it be generally known — without expressly mentioning my authority.

First of all, it will be remembered that Laplace regarded the universe as composed of a number of beings having assignable positions and movements, and ranging in size from the largest celestial bodies down to the lightest atoms. He assumed that all these, whether masses or molecules, whether of finite or of infinitesimal dimensions, are amenable to the same mechanical laws; and this assumption is still regarded as "the axiom on which all modern physics is founded."[2] None the less there are some striking differences in the methods followed in the two cases, *i.e.* according as the masses to be dealt with are of sensible or of insensible dimensions. With sensible masses the physicist's procedure is in the main *abstract*, and any exactness he may attain is attained in this manner. But he at least knows the bodies he is investigating, say the sun or the moon, the bob of a

[1] Cf. above, Lecture I, cp. 17.

[2] J. J. Thomson, *Applications of Dynamics*, p. 1.

pendulum or the screw of a steamship. In dealing with molecules or atoms, on the other hand, such identification and individualisation is impossible. His procedure here, if I may so say, is predominantly *idealistic*. Actual perception is replaced by ideal conception. Moreover, the ideal atoms or molecules are often wholly hypothetical, and when not this — as in chemistry, perhaps — are still rather statistical means or averages than actual existences. Further, the exactness which it is known cannot be affirmed of mechanisms of sensible mass, except after manifold abstractions, is assumed, not unfrequently, to apply *literally* to the hypothetical mechanisms of which atoms and molecules and other ideal conceptions form the working parts. We shall thus have to consider the abstract theory first in itself, next in its application to sensible masses, and lastly in its application to insensible masses.

First, as to the *abstract method*. A few sentences from a standard text-book will make clear what is meant by this. In Thomson and Tait's *Natural Philosophy* the division entitled *Abstract Dynamics* begins as follows: —

"Until we know thoroughly the nature of matter and the forces which produce its motion, it will be utterly impossible to submit to mathematical reasoning the *exact* conditions of any physical questions. . . . Take, for instance, the very simple case of a crowbar employed to move a heavy mass. The accurate mathematical investigation of the action would involve the simultaneous treatment of the motions of every part of bar, fulcrum, and mass raised; but our ignorance of the nature of matter and molecular forces, precludes any such complete treatment of the problem. . . . Hence, the idea of solving, instead of the complete but infinitely transcendent problem, another in reality quite different, but which, while amply simple, obviously leads to practically the same results as the former, so far as concerns . . . the bodies as a whole. . . . Imagine the masses involved to be *perfectly rigid*, that is, incapable of changing form or dimensions. Then the infinite multiplicity of the forces really acting may be left out of consideration." After some further elucidation the writers conclude: "Enough, however, has been said to show, *first*, our utter ignorance as to the true and complete solution of any physical question by the only perfect method, that of the consideration of the circumstances which affect the motion of every portion, separately, of each body concerned; and, *second*, the practically sufficient manner in which practical questions may be attacked by limiting their generality, *the limitations introduced being themselves deduced from experience*."

The method above referred to as 'the only perfect method,'—in which the motions of every particle concerned are taken into account—is obviously the very method that Laplace's imaginary spirit is supposed to apply to the universe. We seem meant to assume that this method is *not* abstract—a very questionable assumption to which we shall be brought back later. Meanwhile, turning to the confessedly abstract method with which the actual physicist has to content himself, let us note in what respects the simple question he actually solves differs from the concrete and really quite different question that is propounded. This refers to a particular crowbar, a particular fūlcrum, and a particular material body to be raised at a particular place and date. Assuming that raising the load at a given place means moving it against the gravitational forces at that place, — though in fact these will not be the only forces concerned, — we shall be told that the answer to the question on this score alone will in general vary for every different place, and even, in general, at every different date. But abstract dynamics knows nothing of places and dates; these are the affair of topography and chronology: it knows only of abstract space, time, and motion, as dealt with by geometry and kinematics. Accurately to ascertain the actual forces existing at any place or time would require precise measurements of a complex kind, and precise measurement in the simplest case is, strictly speaking, an impossibility. Abstract dynamics is a mathematical science and therefore does not measure; there would be an end of all exactness if it did. We should be requested accordingly to *state* what the weight of the load is, or at any rate what it may be taken to be. For the same reason the lengths of the two arms of the lever must be *given*, then the power to be applied can be found. Let us next suppose that the lever is made of lead or of lancewood, and that the load consists of dynamite, sheet-glass, or putty. The exponent of abstract mechanics will object again: You are proposing here millions, nay billions, of problems, instead of one. The properties of the lever as a simple machine being in question, we are entitled to replace the material crowbar by a line of equal length fixed at the point answering to the fulcrum, and to regard it as unalterable in form and dimensions. And as to the load, dynamics can deal only with the mass of that; it does not recognise the qualitative differences of material bodies. "In abstract dynamics"—to quote Maxwell—"matter is considered under no other aspect than as that which can have its motion changed by the application of force. Hence any two bodies are of equal mass if equal forces applied to these bodies produce, in equal times, equal changes of velocity. This is the only definition of equal masses which can be admitted in dynamics, and it is applicable to all material bodies, whatever they may be made of."[1]

The gulf between this final abstraction of 'mass' and the material bodies which it replaces is so great that even the physicists to whom it is due often fail to realise how much they have stripped off. The notion of mass leaves far behind it not merely all the diversities of chemical classification, where iron and carbon, oxygen and chlorine are placed wide apart; not merely the variety of secondary qualities, colour, taste, smell, and the like, whereby sensible objects are commonly described; not merely the physical distinction of solid, liquid, and gaseous states, in one or other of which all material bodies are found. A mass has no chemical nature, no physical properties, not even a weight. Even its relation to space differs from that of sensible bodies. Matter has often been defined as that which can, or that which must, occupy space.

[1] *Matter and Motion*, p. 40.

Whatever these definitions may be worth, they cannot at all events be applied directly to mass as just defined. A mass must have position or it could not be moved, but it may be of finite amount and yet have no size, or it may be of any size whatever. It is true that all bodies of sensible dimensions are found to resist compression, or deformation, or both. But these characteristics are due not to mass, but to forces. Moreover, when such changes in the configuration of a body are under investigation, the body is regarded as a system of mass-elements or mass-points, and these either as continuous or discontinuous, as circumstances may determine. Inasmuch, however, as changes of configuration are conceivable in every material body of finite dimensions, the logical implication is that all such bodies consist of mass-points. Thus the question whether matter is discrete, consisting ultimately of atoms, or is continuous and so indefinitely divisible, is not a question that concerns mass. Indeed, not only may a mass of finite volume be divisible as long as that volume itself is divisible; but even if we suppose ourselves to have reached the geometrical point or limit of spatial divisibility, which has neither parts nor magnitude, this puts no limit to the divisibility of mass. As already said, such a geometrical point may be regarded as the seat of a mass that still has both parts and magnitude. "In certain astronomical investigations," as Maxwell points out, "the planets, and even the sun, may be regarded each as a material particle,"[1] or mass-point. Yet these masses require a very great

[1] *Matter and Motion*, art. vi, p. 11.

number to express them when our customary units of mass are used. On the other hand, "even an atom, when we consider it as capable of rotation, must be regarded as consisting of many material particles" or mass-points — although its total mass in gravitation measure be less than the billionth part of a gramme.

But all this will become plainer, and the extreme abstractions involved in the notion of mass more apparent, if we recur again to its definition, regarding it this time synthetically rather than analytically. It is possible to describe the motions of points or figures and the composition or resolution of such motions in a purely formal manner, just as in geometry their situations and constructions are formally described. In this way kinematics, as the science of abstract motion, covers all the ground implied in change of position or change of speed in any body or system of bodies, so far as such motion involves only pure or abstract space and time. By abstract space and time, it need hardly be said, is meant, as I have already incidentally remarked, the space and time of mathematics, not the variously filled space and time of our concrete perceptual experience. Kinematics is then in the strictest sense a branch of pure mathematics, and not an empirical science. But we pass, it may be supposed, from the mathematical to the real, when, in place of merely describing motion, we ask what is moved and what are the causes of such actual motion. The categories of substance and cause here seem to come upon the scene, and they surely transcend the range of the purely mathematical. But is mass conceived by abstract mechanics as a thing or substance; or is force conceived as a cause? The answer, I think, must be negative to both questions. But deferring the question as to force, it must be noted that mass is by no means synonymous with matter, though sometimes used as if it were. "We must be careful to remember," Maxwell tells us, "that what we sometimes, even in abstract dynamics, call matter, is not that unknown substratum of real bodies against which Berkeley directed his arguments, but something as perfectly intelligible as a straight line or a sphere." "Why, then," he asks, "should we have any change of method when passing from kinematics to abstract dynamics? Why should we find it more difficult to endow moving figures with mass than to endow stationary figures with motion? The bodies we deal with in abstract dynamics are just as completely known to us as the figures in Euclid. They have no properties whatever except those which we explicitly assign to them."[1] In entire accord with this we have the statement of Professor Tait, — all the more impressive because of his well-known hankering after the metaphysical, — that "we do not know and are probably incapable of discovering what matter *is*."[2] Matter as substance is, in short, as rigorously excluded from modern physics as mind, as substance, is banished from modern psychology; indeed, matter is not merely excluded but abused as a 'metaphysical quagmire,' 'fetish,'[3] and the like.

[1] Review of Thomson and Tait's *Natural Philosophy*, in *Nature*, vol. xx, p. 214; also *Scientific Papers*, vol. ii, p. 779.

[2] *Properties of Matter*, p. 14.

[3] Cf. Karl Pearson, *Grammar of Science*, passim.

In dealing with mass, then, we are only dealing with a property; and, since it is a property that varies continuously in quantity, it is one that admits of mathematical treatment. Mass, in short, is but another name for quantity of inertia. By inertia the physicist denotes the fact, or to be strictly accurate I should say the well-grounded inference, that a body, so long as it is left to itself, preserves strictly in respect of motion its *status quo*. We can perfectly well imagine any number of such bodies of the most various sizes and shapes moving severally in all possible directions, and all at different speeds, that of zero speed or rest being one. Referred to some defined origin and axes, their apparent changes of size, shape, relative position after a given interval, as well as their apparent changes of speed, could all be dealt with by kinematics. Such motions, in accordance with Newton's First Law, might be called, perhaps have been called, free, or independent, or unconstrained motions. But this is not all that kinematics could do. We might arbitrarily assign to any or all the bodies under contemplation any deviations from uniform rectilinear motion or from rest; and the resulting positions after a given interval might still be found as before. Such deviation from uniform rectilinear motion or from rest is, of course, in actual fact the rule; and the kinematical problems of abstract dynamics — if I might so call them — differ from such arbitrary problems only in not being arbitrary. "The new idea appropriate to dynamics (then) is" — I quote Maxwell — "that the motions of bodies are not independent of each other, but that, under certain conditions, dynamical transactions take place between two bodies whereby the motions of both bodies are affected."[1]

Now one of these conditions is that the said transactions between two bodies — as Maxwell picturesquely calls them — are in no ways affected by, and in no ways affect, other dynamical transactions which either or both the bodies may have with other bodies. In a word, the results of all such transactions are *additive.* All the principles involved may therefore be learnt by considering such a transaction in a single case. Another condition is that such transaction between two bodies takes place along the line joining them; also, that the changes of motion or the accelerations of each body along this line, in which the said transaction or mutual stress consists, are in opposite directions. But how far is each to shunt from its original direction, how much is each to alter its original speed, that is to say, what share in the whole transaction is each to take? *The answer to this question gives the meaning of mass.* To each body a number is to be assigned, such that the changes of their motion are inversely proportional to these numbers. Such number answers to the mass of the body to which it belongs. Its determination, of course, in any real case involves measurement, and is the business, not of abstract dynamics, but of experimental physics. The actual number again depends on the standard employed, but, once so determined, by dynamical transaction with the standard, it is determined once for all for every other dynamical transaction with other masses numbered according to the

[1] *Nature, l.c.*

same unit. The appropriateness of defining mass as quantity of inertia, *i.e.* as the measure of that tendency to persistence of the motor *status quo* which preceded the particular dynamical transaction under investigation, is thus evident. For the greater the mass, the less in any given case the change of motion that ensues; the less the mass the greater the change of motion — kinematically estimated, of course. Thus, if the mass of one of the two bodies is infinite, its kinematic circumstances are unaltered; if the mass of one be zero, that of the other, however small, undergoes no acceleration; where both are equal, the accelerations of both are equal; and so for every other case. So far then from falling under the category of substance, a mass as it occurs in abstract dynamics is but a coefficient affecting the value of the acceleration to which it is affixed. True the phrase "*mass of a body*" is constantly recurring; but then the body, apart from the mass, is but a moving point or figure.

There still remains the correlative term Force. How, it may be asked, can the bodies of abstract dynamics be conceived as merely geometrical figures moving according to rule, if they are collectively endowed with all the forces of nature: gravitation, light, heat, electricity, chemical attraction, etc.? What are these if they are not the active properties of material bodies? The investigation of the nature of matter or of the properties of real bodies, we shall be told, is entirely the business of experimental physics; abstract dynamics takes account of no properties but those expressed by its definitions. But by definition a body is endowed with no essential properties but mass and mobility. Force, as understood by dynamics, cannot then be an inherent and permanent property of any given body, dynamically considered. On the contrary, no mass, though infinite, has any force by itself. A force in the dynamical sense cannot appear till there are two masses in dynamical relation, and then there will be two equal and opposite forces, let the masses differ as much as they may. A force is but the name for a mass-acceleration, *i.e.* for either side of the dynamical transaction between two bodies, which we have already considered; and a moment's recurrence to that transaction will make the purely mathematical character of such forces plain. Instead of the moving geometrical point of kinematics, we have in dynamics a mass-point in motion. This mass-motion for a given direction is called momentum; momentum being the product of the number of units of mass into the number of units of speed. It remains, so long as the body is left to itself, a constant quantity. When two masses are said to interact, the momentum of each changes, and the rate of this change for one of the bodies is called the moving force on that body; this again is a quantity, the product, as said, of mass into acceleration. In short, the old qualitative definition of force as "whatever changes or tends to change the motion of a body" is discarded by modern dynamics, which professes to leave the question of the causes of such change entirely aside. *Force for it means simply the direction in which, and the rate at which, this change takes place.* It answers, says Kirchhoff, in mathematical language to the second differential coefficient of the distance as a function of the time; is, as Tait puts it, no more an objective entity than say five per cent per annum is a sum of money.[1]

How completely the theory of mechanics has divested itself of the conceptions of substance and cause, in assuming its present strictly mathematical form, is brought home to us by one striking fact; the fact, I mean, that mass and force, in which these categories are supposed to be implied, are but dependent variables in certain general equations. In $7+5=12$ or $\tan 45^\circ=1$, we cannot say that one side of these equations is more than the other effect or consequent, that other being the cause or essence whence it proceeds. It would be equally arbitrary to attempt any such distinction when we have the equation $mv=ft$, or $ms=ft^2/2$ or $fs=mv^2/2$. In these, the fundamental equations of dynamics, we have four quantities so connected, that if any three are known the fourth can be found. In this respect one term is no more real than another, and the dependence is not temporal or causal or teleological, but mathematical simply. The sole use of such equations, it is contended, is "to *describe* in the exactest and simplest manner such motions as occur in nature." So Kirchhoff defined the object of mathematical physics in his universally lauded text-book, and his definition has recently been made the motto of a manifesto on the part of Professor Mach. "It is said," Mach remarks, "description leaves the sense of causality unsatisfied. In fact, many imagine they understand motions better when they picture to themselves pulling forces, and yet the *accelerations*, the

[1] Cf. Tait on *Force*, *Nature*, vol. xvii, p. 459.

facts, accomplish more, without superfluous additions. I hope that the science of the future will discard the idea of cause and effect, as being formally obscure; and in my feeling that these ideas contain a strong tincture of fetishism, I am certainly not alone."[1]

I am quite aware that the elimination from natural science, of this so-called fetishism, which the categories of substance and cause are supposed to involve, has been gradual.[2] But the history of mechanics shews conclusively that there at any rate this process of elimination has been steady, and now at length seems to be complete. The full significance of this deanthropomorphic tendency of science it will be best to defer, along with other epistemological reflections, till we have reached the end of this survey of the cardinal doctrines of modern science, which we have but just commenced. At this stage I will only venture the remark that those who seek to oppose this tendency—as Wundt and still more Sigwart, for example, seem to do—appear rather to mistake the issue. It is not a question of divesting the human mind of its most fundamental conceptions; it is simply a question of method and expediency, the propriety, in a word, of dividing natural science from natural philosophy. No doubt many of those who insist on this separation are privately of opinion, as we have seen,

[1] *Popular Scientific Lectures*, Eng. trans., p. 253.

[2] Even in the time of Newton forces were regarded as powers inherent in substances. Their *effects* could be measured, but not the forces themselves. Still earlier the *remora* or *echineis*, though but a "little fish," was credited with the power of stopping a ship by merely adhering to it. Cf. Whewell, *History of Inductive Sciences*, 3rd edition, vol. i, p. 189.

that natural science will make a whole of knowledge by itself. But in so thinking they are only playing the amateur philosopher. Such a declaration is no part of their business as scientific experts. As Mr. Bradley roundly puts it: "When Phenomenalism loses its head and, becoming blatant, steps forward as a theory of first principles, then it is really not respectable. The best that can be said of its pretensions is that they are ridiculous."[1] The sharper the division of labour, the more fragmentary becomes the contribution of each separate worker; but the more perfect also the finished production of their joint organisation. The 'ragged edges' of scientific knowledge ought to become more apparent the more strictly scientific they are; and the more defined these ragged edges are, the more effectively can philosophy enter upon the work it aspires to do, of articulating or healing those sutures, of rounding off and unifying the whole.

No wonder Laplace could dispense with the hypothesis of a Deity, if his celestial mechanics turn out to be so abstract as to exclude the categories of substance and cause. A mathematical formula does not change its essential character by increasing in length and complexity. If the validity of an equation is by its very definition confined to what is mathematical, if it is only tentatively and approximately applicable to what is real, Laplace's world formula must be like the rest. On this question of the relation of abstract dynamics to actual phenomena, I propose to enter in the next lecture.

[1] *Appearance and Reality*, p. 126.

LECTURE III

RELATION OF ABSTRACT DYNAMICS TO ACTUAL PHENOMENA

The characteristics of Abstract Dynamics recapitulated.

The question raised : How far, and in what sense, this science can be applied to actual phenomena. *This problem illustrated from Newton's treatment of Space, Time, Motion, as* (1) *absolute ;* (2) *relative.*

Bearing of this distinction on the attempt to determine an actual case of the first law of motion. Various proposals considered. The question of absolute rotation especially instructive. Mach's criticisms reveal the indefinite complexity of 'real cases.'

The mechanical theory is thus divided against itself: it cannot be at once rigorously exact and adequately real. The Kirchhoff School abandon the attempt "to penetrate to the mechanism of nature," and see in mechanics only an instrument for 'approximate description.' Unconditional mechanical statements concerning the real world appear so far unwarrantable.

One of these specially discussed: the Conservation of Mass. *Mr. Herbert Spencer's 'short and easy method' found wide of the mark. This doctrine, like other mechanical doctrines, justified mainly by its simplicity.*

WE resume to-day the attempt to estimate the validity and the scope of the mechanical theory of the universe. To understand this we have had first of all to inquire into the precise import of the science of abstract mechanics or dynamics, on which that theory is avowedly founded. We have accepted the declaration of mathematical physicists in the present day that it is not the province of mechanical theory to explain phenomena by means of natural forces, but only to describe completely in the simplest possible manner, such motions as occur in nature.[1] We appreciate most readily the distinctive character of pure mechanics, as thus defined, if we approach it from the side of kinematics. Kinematics is held to suffice for the description of any actual or possible motion of bodies, regarded as moving figures of constant or varying shape. If there are some motions too complex for kinematic treatment in the present state of that science, the defect is one that mechanics can do nothing to remove. But "the motions that occur in nature" are frequently, and, it is supposed, are always, mutually dependent. As to the character of this dependence, the most various hypotheses might be — indeed have been — formed; and when such hypotheses are sufficiently definite, as regards their space and time elements, their kinematical consequences can be deduced. The kinematical problems thereby entailed *might* be appalling in comparison with those required by the simple assumptions

[1] It may be objected that such 'simplest possible description' is itself explanation, that in fact explanation is merely resolving the complex into the simple, and assimilating the less known to the better known. I admit this fully. But experience is not restricted to the range of exact science, and so far it is true that a fact is not fully explained if its cause is unknown. (Cf. below, Lecture XIX.) Precisely in this lay the difficulty for such men as Huygens, Leibnitz, and Bernoulli of Newton's theory of gravitation. Newton only professed to 'describe,' but, as Lange tersely puts it: "These men could not separate the mathematics from the physics, and physically the doctrine of Newton was for them inconceivable." And so it has remained till this day, although people are now accustomed to regard Newton's descriptive conception as if it were itself a physical cause.

to which, after many trials, Galileo, Huygens, and Newton, the founders of modern dynamics, were led. By means of the conception of mass the notion of quantity of motion, or momentum, was made definite by Newton, and the so-called laws or axioms concerning momentum formulated. According to these the rate at which their momentum changes, when two masses are in the state of mutual stress, is always equal in amount, their motions taking place in opposite directions along the line joining them, the result being that the momentum of their common centre of mass remains unchanged.

Nothing could be more sublimely simple, especially when it is remembered that these axioms involve the so-called parallelogram of forces; imply, that is, that the mutual accelerations of any two masses are uninfluenced by the presence of a third mass. Such is abstract dynamics; and, regarded from within, its exactness is as impressive as its simplicity. Not only is it clear of such 'bottomless quagmires' as substantiality and causality, conceptions which no *science* has ever yet adjusted to facts; but as 'rational mechanics'[1] it is clear, too, of all induction and all experiment, resting wholly, as truly as any formal science does, on its own fundamental definitions and axioms. The only space or time or motion that it knows is what Newton called absolute, true, and mathematical, and sharply distinguished from the relative spaces, times, and movements of our perceptual experience.

How far, and in what sense, this pure mechanical science can be applied in the phenomenal world is now for us the

[1] Cf. Newton's Preface to the *Principia*.

vital question. Unhappily the authorised teachers of physics seem only recently to have waked up to the possibility of such a question at all. The only 'applied mechanics' they seem aware of is that of the mechanician and the engineer. While admitting readily that the astronomer applies geometry and trigonometry in his investigations, they talk as if he were entirely in the region of pure theory as soon as he proceeds to discuss celestial movements. Newton at all events knew better than this, even if he realised the difficulty of the transition less than many now do. Let me quote a few sentences from the *Principia* in illustration.[1] First, as to *time:* "Absolute, true, and mathematical time, in itself, and from its own nature, flows equally, without relation to anything external; and by another name is called Duration. . . . The natural days are truly unequal, though they are commonly considered as equal and used for a measure of time. Astronomers correct this inequality that they may measure the celestial motions by a more accurate time. It may be that there is no equable motion, whereby time may be accurately measured. All motions may be accelerated and retarded; but the flowing of absolute time is liable to no change. Duration . . . remains the same, whether motions are swift or slow or none at all: therefore this duration is properly distinguished from its sensible measures; and from them it is collected by means of an astronomical equation."

Again, as to *space:* "Absolute space, in its own nature, without relation to anything external, remains

[1] Cf. Pemberton's translation, pp. 10 ff.

always similar and immovable." "For the primary places of things to be moved is absurd. These are therefore absolute places; and translations only out of these are absolute motions. But, because the parts of space cannot be seen, or distinguished from one another by our senses, therefore in their stead we use sensible measures . . . and that without any inconvenience in common affairs: but in philosophical disquisitions, we must abstract from the senses. For it may be that no body is really at rest, to which the places and motions of others may be referred. . . . It is possible that in the regions of the fixed stars or far beyond them, there may be some body absolutely at rest; but yet [it is] impossible to know from the position of bodies with respect to one another in our regions, whether any of them do keep the same position to that remote body or no. It follows [therefore] that absolute rest cannot be determined from the position of bodies with respect to each other in our regions."

Lastly, as to *motion:* "Absolute motion is the translation of a body from absolute place to absolute place; and relative motion is the translation from relative place to relative place." "If a place is moved, whatever is placed therein is moved along with it. . . . Therefore all motions which are made from places in motion, are only *parts* of entire and absolute motions: and every entire motion is composed of the motion of the body out of its first place, and of the motion of this place out of its place, and so on, until we come to some immovable place, as in the example of the sailor before mentioned [who was supposed to move relatively to his ship which moved relatively to the earth, which in turn moved relatively to the sun, and so on and on]. Wherefore entire and absolute motions can be no otherwise determined than by immovable places. . . . It is indeed a matter of great difficulty to discover and effectually to distinguish the true motions of particular bodies from the apparent: because the parts of that immovable space, in which motions are truly performed, do not come under the observation of our senses. Yet the case is not *altogether desperate;* for arguments may be brought, partly from the apparent motions, which are the differences of the true motions; partly from the forces, which are the causes and effects of true motions."

One can readily gather from statements like these that Newton saw no difficulty in working out problems in which the time should flow at a constant rate, and in which motion from absolute place to absolute place was at once and effectually determined. The position of mechanical theory is in this respect precisely on a par with that of geometry. The description of the circle, say, is easy and exact, but accurately to describe the figure of any real object is an impossibility. So it is with the fundamental quantities concerned in physics.

It is impossible to find in nature or artificially to construct an accurate timekeeper. The physicist simply has to collect the true time from its 'sensible measures,' to use Newton's phrase, as nearly as he can. Experience provides us with innumerable instances in which processes seemingly identical in character and severally independent, are again and again repeated in such wise that the number of repetitions of one kind of process is found to bear an approximately constant ratio to the number of repetitions of another and contemporaneous series. The solar day, the lunar month, the solar year, so far as we may regard them as independent events, are instances of such isochronous series of the natural sort; the periods of waves of light or of waves of sound are other instances; while the vibrations of a given spring or a given pendulum are cases of artificial isochronous events. The comparison of a number of such series — aided by dynamical reasoning, whereby certain disturbances can be ascertained and corrected, and aided again by the theory of probability in eliminating errors of observation — results in the attainment, not of a measure flowing equably without regard to anything external, but in the best mean value possible in our restricted circumstances. Between such mean time and absolute time there is a difference, that is certain; and that difference is, for the mechanical theory, of the nature of error or defect. It is immaterial to the question we have in hand whether absolute time is also real or is ideal only. It is at least ideal, and the fact that the physicist has to leave this ideal behind him when he proceeds to apply abstract dynamics to natural phenomena is the fact to be noted.

Turning to space, the same fact meets us again. Instead of the immovable space, the fixed axes, the primary places of mathematical theory, we have that indefinite regress from relative place to relative place, which renders the attempt to ascertain the so-called true motions of particular bodies, as Newton allows, "well-nigh desperate." Consider, for example, a case falling under the first law of motion. According to this law the motion of a body free from external forces is uniform in magnitude and direction. The mathematician has no trouble with this. He can always specify the axes to which he refers, and plot out diagrams of velocity in his paper space. But when we pass to empirically given space, where is the place to which the direction of a body moving under the action of no forces is referred? "A number of writers," says Professor MacGregor in a recent article, "have attacked this problem, and left it only half solved."[1] Newton's forlorn suggestion that possibly in the region of the fixed stars, or far beyond, there may be a body absolutely at rest, to which the positions and motions of others may be referred, has been revived. In favour of assuming this fictitious Body Alpha, as it has been called, it is urged that such a body provides an escape, in thought at all events, from the hopeless confusion of relative motions to which there is no end.[2] But *ideally* this Body Alpha is not wanted, and *practically* it is useless. Another and less chimerical method that has found more favour begins, not by asking for a body absolutely at rest as a fundamental point of orientation, but by asking for an "inertial system." To constitute such a system it suffices to have three particles projected at the same instant from one position, and each left free to move, uninfluenced by force. Then, provided they do not all move in one straight line, it is geometrically possible to find axes, referred to which they will all three

[1] *Hypotheses of Dynamics*, *Phil. Mag.*, 1893, vol. 36, p. 237.

[2] Cf. Sigwart, *Logic*, § 88, 8; and Riehl, *Der philosophische Kriticismus*, Bd. II. i. pp. 92 ff.

move in straight lines. Referred to such a system, the path of any fourth body moving free from force will be a straight line.[1] But this again is obviously theoretical, and so far superfluous. Practically it is as impossible to ascertain that a body is absolutely free from forces as it is to ascertain its direction relatively to the Body Alpha, the presumption being indeed that no such body, unless it be the universe as a whole, exists. Yet a third method has been proposed of answering the question: Relatively to what, is a body free from constraint moving uniformly in a straight line? The answer according to this method, which has been adopted by Professor Tait, is, "Relatively to any set of lines drawn in a rigid body of finite dimensions, which is not acted on by force, and which has no rotation."[2] Here again it may be objected that it is impossible to find such a body, for if the universe is a single mechanical system, there is no such body to find.

But none the less this method brings to our notice a topic keenly canvassed nowadays among physicists, which is of extreme interest; so that I trust I may be pardoned for meddling with it. Newton believed that he had shewn, first by experiment, and then by theoretical reasoning, that "there is," as he puts it, "only one real circular motion of any revolving body . . . whereas relative motions in one and the same body are innumerable." Thus, if two bodies in an immeasurable void were found to approach, there would be no means of determining which was moving. But if the two bodies were con-

[1] Cf. L. Lange, *Die geschichtliche Entwicklung des Bewegungsbegriffes*, 1886, p. 139.

[2] *Properties of Matter*, p. 92.

nected by a cord, it would be possible, though their distance remained unchanged, to determine whether they were revolving or not. To settle this question it would be sufficient to ascertain the presence or absence of tension in the cord. Accordingly it is argued, as by Professor Tait, that a body *not* rotating will provide us with fixed directions in space, constitute a sort of absolute compass, so to say; and by the help of Newton's physical test it can be ascertained whether a body has rotation or not. Here, then, we seem to have something absolute, an exception to the supposed invariable relativity of everything phenomenal. But so far we have been given only a purely hypothetical case — a single system in an immense void. Newton's actual experiment consisted in rotating a bucket of water by strongly twisting a cord suspending it, so as to make the bucket spin rapidly. At first, when the bucket alone rotates, the surface of the water remains flat, although relatively to the bucket it is not at rest; whereas, by the time the water revolves along with the bucket its surface has become concave, thereby evidencing "real circular motion," to use Newton's phrase, notwithstanding that the bucket and the water by this time are at rest relatively to each other. Finally, when the bucket has ceased to revolve, the surface of the water continues concave some while longer, because "its endeavour to recede" from the axis has not yet ceased. "Therefore," says Newton, "this endeavour does not depend upon the translation of the water in respect of the ambient bodies, nor can true circular motion be described by such translation." In other words, as Kant remarks, "a motion which is a change of external relation in space can be given empirically, although this space itself is not empirically given, and is no object of experience — a paradox deserving to be solved." Kant's own solution is of interest in its way, but it does not help us much, for it leaves the paradox in the main as he found it. But I will ask your attention instead to the much more trenchant criticism of Mach, as this will serve to illustrate the epistemological difference between abstract science and its empirical application, which is our immediate theme.

First of all let us note the difference between Newton's theoretical instance and his experimental one. In the purely hypothetical case we imagine a single mass system in an immense void, and it is shewn under what circumstances, provided the Newtonian laws of motion are assumed, the rotation of such a system could be demonstrated. In the real case, which is meant to verify this deduction, we are confined entirely to experimental methods. But now in this case, over and above the rotating mass of water, we have not only the mass of the bucket, but we have also the masses of the earth, of the rest of the solar system, and of the so-called fixed stars. Now, says Professor Mach, "Newton's experiment . . . only shews us that the rotation of the water relative to the sides of the bucket occasions no perceptible centrifugal forces, but that such forces *are* occasioned, when the water rotates relatively to the masses of the earth and the other heavenly bodies."[1] Experimental canons then at once suggest

[1] *Die Mechanik in ihrer Entwicklung*, 2te Aufl., pp. 216 f. There is now an English translation of this most interesting book.

two further inquiries: Might not the rotation relative to the bucket have some effect if the sides of the bucket were enormously increased in thickness? Or again — allowing for the moment that the proposition is not absurd, at least not kinematically absurd — supposing the bucket to be fixed and the whole choir of heaven to circle round it, would there then be no sign of rotation in the water? Such experiments being impracticable — for, as Mach well says, "the universe is not presented to us twice, first with the earth at rest and then with the earth rotating" — we are left to content ourselves, as best we can, with this result: that a body with so-called absolute rotation is a body rotating relatively to the fixed stars; and that a body without rotation means a body directionally at rest, *not absolutely*, but relatively to the fixed stars.

Returning now for a moment to Newton's hypothetical case, it is obvious that a physicist actually confined to such a system, before he could begin experimentally to apply or to verify the Newtonian laws of motion, would find himself face to face with the very difficulties we have considered. Positions and directions must be independently determined before dynamical investigations are begun. To assume the laws of motion in order to fix directions and then to use these directions in order to establish the laws would be obviously fallacious. From such a logical circle *abstract* dynamics is free, because the physicist has here the complete command of ideal space, as is shewn by his diagrams on paper; and because he has not to prove the laws of motion, but merely to deduce their theoretical consequences. Newton's absolute rotation is then, like his absolute time and absolute space, not real but ideal, not sensibly or empirically given but intellectually conceived or constructed, not ectypal but archetypal, as Locke says of all purely mathematical ideas.

This becomes clearer, if we consider the difference between the two cases from another side. The hypothetical case is that of a finite system in an immense void; all the rest of the universe is supposed to be eliminated. In the real world we may ignore, but we cannot exclude. Thus, as already said, it is allowed that—except by accident—there is probably no body in the state described in the first law of motion, in fact, if the master generalisation of physics, the law of universal gravitation, is to be accepted, how can any particle of matter "be left to itself"? By a free particle, or a particle left to itself can only be meant a particle at an infinite distance from any other particle, and in this sense accordingly writers on abstract dynamics sometimes define the phrase. But if we could come across such a particle in actual experience, it is obvious that nothing could be said about it; spatial perception of any kind would necessarily be absent in such circumstances. In dealing with the actual world, however, the facts that meet us first are those to which Newton's second and third laws apply, and the law of inertia becomes but a special case of these. Setting out from these laws, then, instead of attempting to affirm anything concerning the movement of a particle alone in absolute space, it seems to me as a mere question of scientific taste and logic better to proceed in Mach's fashion. Instead of saying that a particle moves without acceleration in space, Mach would say that the mean acceleration of such particle relatively to the other particles in the universe, or in a sufficient portion of the universe, is zero.[1]

As it is obviously impossible to complete the summation required to ascertain this mean exactly, such a statement has the advantage of keeping prominent the approximate character of references to the directions of certain stars as *fixed* directions. The reference to fixed terrestrial objects, which sufficed for such observations as led Galileo at first to formulate the law of inertia, is now replaced by this reference to fixed stars; but even this direction is known to change in the course of ages. Another advantage of Mach's more concrete statement, then, is that it impresses us, as he remarks, with the very complicated character of just those mechanical laws that appear the simplest. Suggested by incomplete experiences in the first instance, they lose the exactness of mathematical theory when we proceed to apply them to experience again. The manifold particulars left out of account in our abstract simplification are still there on our return to confront us anew. The insight that a pure theory has given may enable us to deal with them more effectually; it cannot justify us in ignoring their existence. Now by good fortune, not from any necessity in the constitution of things, it is found that within certain limits of exactness many of these particulars of experience are so similar, that any one appears to suffice for the rest. One result of this apparent multiplicity of

[1] But see the article by Professor MacGregor quoted above.

identicals is that, seeming to be independent of any one, we presently suppose ourselves independent of all; when to be absolutely exact we are independent of none. In applying the law of inertia to terrestrial bodies, for example, there are innumerable landmarks from which to estimate direction; if one or more become unsteady or disappear, there are still plenty of others left. So with celestial objects; if one fixed star should some day "pale its feeble light" or be found careering across the sky, there are still multitudes remaining to keep their accustomed stations. Now, it is our familiarity from time immemorial with this plenitude of possibilities that leads us to convert these several singular contingencies into a collective contingency. We then assume that, as we are independent of any one empirically marked position in space, we are independent of all. In other words, the absolute space of abstract conception is supposed to underlie the empirical space that we perceive. But now imagine, as Mach suggests, that the earth were the scene of incessant earthquakes or that the stars behaved like a swarm of flies: how should we apply the law of inertia then? Well, but to those who mean seriously to handle the universe as a mere problem in abstract dynamics we must reply that the earth *is* the scene of incessant convulsions and the fixed stars *are* like a swarm of bees. The costliness of the devices to eliminate terrestrial oscillations in certain attempts at experimental precision and the elaborate calculations to unravel the 'proper motions' of the less distant stars are plain evidence of the truth of this seemingly extravagant statement.

It would seem then that *all* bodies may be really implicated in every case of movement observing the law of inertia; not *one* only, as the abstract theory assumes. What a single body would be or do if it were not for other bodies, no one can say. Unless indeed they are prepared with Stallo to say boldly, it would be nothing and therefore could do nothing. "A body," he says, "cannot survive the system of relations in which alone it has its being; its *presence* or *position* in space is no more possible without reference to other bodies than its *change of position* or *presence* is possible without such reference. . . . All properties of a body which constitute the elements of its distinguishable presence in space are in their nature relations and imply terms beyond the body itself."[1] In abstract theory, then, we may introduce first one particle and then another, each moving in given directions in absolute space; and we may talk of their speed as measured by absolute time flowing equably without relation to anything else. But, in reality, nothing of this kind is accessible to us.

It is easy to see that the mechanical theory is here divided against itself, and in this state cannot stand. Experience compels it to admit the thorough-going interdependence of all bodies, while mathematics tempts it to suppose that it is possible to deal with bodies independently and apart. The bodies which mathematics would regard as isolated wholes are but undetermined fragments of what is really indivisible, abstract aspects that never exist alone. On the one side is the ideal simplicity and completeness of a mathematical creation; on the other an illimitable complexity of relations without beginning,

[1] *Concepts and Theories of Modern Physics*, p. 200.

without middle, and without end. Now I presume nobody will blame the physicist for insisting on the relativity of all motion, the relativity of all time-measures, which practically depend on motion, or the relativity of all determinations of mass or inertia. But we have a right to demand logical consistency: if he abjure absolute terms he must abjure absolute statements. He must not confound his descriptive apparatus with the actual phenomena it is devised to describe. The apparatus consists, in general, as we have seen, of absolute time, that is, an independent variable flowing at a constant rate; of absolute motions, that is, motions referred to axes completely defined and thought of as fixed; of bodies that by definition are masses and only masses, absolutely determinate and unchangeable, and constituting together a mechanical system that is independent and complete. Of this general form of apparatus there may be several varieties, but that will be accounted the best which affords the simplest and completest description of actual movements. We cannot be sure that there is any *a priori* necessity about the particular mechanical principles of Galileo and Newton; from other fundamental definitions consequences equally exact might be deduced. As this is an assertion that to many may seem unwarranted, let me hasten to say that I do not make it without good authority; I will quote one such out of many. In an essay on the *Methods of Theoretical Physics*, Boltzmann, referring with approval to the changes introduced by Kirchhoff, thus proceeds: "Whether, with Kepler, the form of the orbit of a planet and the velocity at each point is defined, or with

Newton, the force at each point, both are really only different methods of describing the facts; and Newton's merit is only the discovery that the description of the motion of the celestial bodies is especially simple if the second differential of their coördinates in respect of time is given."[1] In either case, and in every case, then, we have only mathematical description. "The whole difficulty of philosophy," said Newton, in the Preface to his *Principia*, "seems to consist in investigating the powers of Nature from the phenomena of motion." Many of his successors have abandoned the enterprise. To quote Boltzmann again: "The view [has] gained ground that it cannot be the object of theory [*i.e.* of science] to penetrate the mechanism of Nature, but that, merely starting from the simplest assumptions (that certain magnitudes are linear or other elementary functions), to establish equations as elementary as possible which enable the natural phenomena to be calculated with the closest approximation." Equations, not explanations, approximation, not finality, and the simplest method the best: in such wise has the modern science of dynamics narrowed its scope. And the criterion of simplicity, it must be remembered, is in the main subjective, not objective. Our limited capacities make economy a consideration. But for such limitation, indeed, it is difficult to see why we should cumber ourselves with a descriptive apparatus of any sort. It is surely then a thoughtless prejudice to forget that the capacity to calculate and compute — though, as Laplace boasts, it renders the human species superior to the animals, and is

[1] *Philosophical Magazine*, 1893, vol. 36, p. 40.

the foundation of our glory — is also still, like apparatus generally, essentially a mark of limited powers. Regarded in this light it becomes very much a question whether the Newtonian scheme is even the simplest; indeed, other schemes, professedly simpler — and what, if true, is of greater moment, *more comprehensive* — are already in the air. If human capacities are limited, they are not stationary. As Kirchhoff remarks: "A description of certain phenomena, though it be indubitably the simplest we can now give, may in the further progress of science be superseded by another simpler still. Of such like changes the past history of mechanics furnishes instances in plenty."[1] Still this question of comparative simplicity does not concern us save as it may serve to impress two points. First, the difference between the means of description, "the conceptual shorthand," as Professor Karl Pearson happily styles it, and the perceptual realities it is devised to symbolise and summarise. Secondly, the absence of finality. A possible form of description is not enough, it must be shewn to be the only one possible, the only one that the phenomena themselves allow, before it can be held to have passed out of the region of hypothesis into that of *objective truth*.[2]

The conclusion then to which we are led is plain. *The application of abstract mechanics to real bodies is throughout hypothetical*, and absolute or unconditional mechanical statements concerning the real world are therefore unwarrantable. There are no processes in

[1] *Vorlesungen über mathematische Physik*, p. 1.

[2] Cf. Helmholtz, *Erhaltung der Kraft*, p. 7.

the real world that are certainly entirely mechanical, mechanical in the sense, I mean, of those movements of sensible masses from which Galileo and Newton inductively inferred their well-known laws. The thermal, chemical, electrical, magnetic, and other processes that as a rule not only accompany but modify such mechanical movements *may* admit of complete and simple description in purely mechanical terms. But there is no necessity that they should. Newton saw reason to hope for it, however. In the Preface to his *Principia*, he justifies its title as *Mathematical Principles of Natural Philosophy* by referring to the motions of the planets, the comets, the moon, and the sea as deduced from gravitational forces by propositions that are mathematical. He then adds, "I wish we could derive the other phenomena of nature from mechanical principles by the same kind of argument. . . . But I hope that the principles here established will afford some light either to this, or some more perfect method of philosophy." It is to this subject that we must pass in the next two lectures, and we shall then have an opportunity of inquiring which of Newton's alternative hopes is the more nearly realised: the resolution of natural phenomena that are not obviously mechanical into mechanisms, or the advent of some more perfect method embracing both. But either way our main conclusion will, I believe, still remain good.

There is one absolute statement frequently advanced by modern physicists that flagrantly transgresses the limits of a purely descriptive science, the statement, I mean, that the mass of the universe is a definite and unchangeable quantity. Such partial and approximate evidence as experience affords in favour of such a doctrine seems to be derived ultimately from the facts of gravitation. Astronomical observations of planetary motions and chemical measurements with the balance justify the working hypothesis that such sensible masses as we know are constant within the limits of our experience and unalterable by any means in our power. Thus has been suggested the addition to abstract dynamics of a principle not explicitly formulated by Galileo or Newton, that, namely, of the Conservation of Mass, as it is technically called. If the mass-values of bodies were assumed to vary in some regular manner with the time, with the size or proximity of neighbouring systems, or the like, the procedure of abstract mechanics would be more complicated than it proves to be on the simpler hypothesis of the constancy of such mass-values. But though actual facts conform to such an assumption, there is no necessity about it. Still less is there any justification for converting this principle of mass-conservation into an assertion concerning the mass of the universe either in respect of its quantity or its constancy. The epistemological character of mathematical mechanics as a purely descriptive apparatus would exclude these, as well as other real affirmations, from its scope. It would be as reasonable to expect from arithmetic a census of the separate bodies in the universe as to look to pure mechanics for an assurance that the mass of the universe is, as Helmholtz would have us regard it, an eternally unchangeable quantity. If there are any grounds for such a position at all, those grounds must lie either in *a posteriori* inferences from experience, that can never be more than probable, or in *a priori* reasoning of a non-mathematical kind.

But before *a priori* considerations can be brought to bear on such a point, mass must be identified with matter, and matter with substance. And this is precisely what we find in the plausible and summary argument of Mr. Spencer's *First Principles*. His crucial experimental proof is just that constancy of mass, gravitationally measured, which I have already mentioned. For, after citing several trivial instances, he clenches them with the remark: "Not, however, until the rise of quantitative chemistry, could the conclusion suggested by such experiences be reduced to a certainty."[1] Spite of this very restricted evidence for the conservation of mass as a simple and useful working hypothesis, we find Mr. Spencer concluding that "the form of our thought renders it impossible for us to have experience of Matter passing into non-existence, . . . that hence the indestructibility of Matter is in strictness an *a priori* truth"; albeit the 'pseudo-thinking of undisciplined minds' is ever leading them mistakenly to suppose they can really think 'the absolutely unthinkable.' Now the question is not at all whether we can or cannot conceive the universe to arise out of, or pass into, nothing; but simply what justification there may be for a certain absolute statement concerning that dynamical phenomenon we describe by help of the conception of Mass. When Mr. Spencer or some one else shall have shewn that what exists must exist as matter or not exist at all, and that

[1] *First Principles*, p. 173.

all matter is necessarily ponderable matter, then, but not before, the old maxim, *Ex nihilo nihil fit*, and the appeal to the balance will be relevant to the question.

Quantity of mass is not necessarily identical with quantity of matter; and indeed, it seems obvious that, till matter is defined qualitatively, quantitive statements concerning it must be altogether precarious. Meanwhile, the prospects of a scientific definition of matter get more and more remote. The severely exact physicist of the Kirchhoff school, as we have seen, avoids the whole of this subject with disdain; while others with powerful scientific imagination like Faraday or Maxwell or Lord Kelvin, who pursue it eagerly, find themselves eluded in turn, and end, as Boltzmann says, by talking in parables.[1] Yet such parables and analogies are of inestimable value, if only as a protest against the confident dogmatism of which Mr. Spencer is such a master. Consider, for example, Lord Kelvin's well-known vortex-atom theory of ponderable matter. According to his ideal

[1] Roger Cotes begins his Preface to the *Principia* by reducing natural philosophers to three classes: first, the Aristotelians, who attribute specific and occult qualities to things, and last, the experimentalists, who invent no hypotheses, among whom, of course, he places his 'most celebrated author.' The second reject the substantial forms of the peripatetics and lay down the principle that all matter is homogeneous. "But when," he continues, "they assume to themselves a liberty of supposing at pleasure unknown figures and magnitudes, uncertain situations and motions of the parts; and moreover of supposing occult fluids, which freely pervade the pores of bodies, endowed with an all-powerful subtility, and agitated with occult motions; they then descend to visions, and neglect the true constitution of things. . . . Although they afterward proceed with the greatest accuracy from those principles [they] may be said to compose a fable, elegant, perhaps, and pleasing to the imagination, but still it is a fable."

presentation of it we are to imagine a perfect, *i.e.* absolutely frictionless fluid; then the rotational motion of portions of this fluid are what we know as ponderable matter; while the movements of these through the fluid are what we know as moving masses. In other words, our phenomenal matter is reduced to 'non-matter in motion.' This brilliant hypothesis (which has been accounted deserving of careful and minute attention by many of our leading physicists), suffices, even as it stands, to suggest what removes there may be between our physical experiences and anything that *must* be conserved because its non-conservation is *a priori* inconceivable. But instead of taking this hypothesis as it stands, let us suppose, as the writers of the *Unseen Universe* do, that its ideal rigour is somewhat abated. Vortex rings in an *absolutely* perfect fluid would remain self-identical and undiminished forever; vortex rings in an indefinitely perfect fluid would so remain, not forever, but indefinitely long. But *per contra*, vortex rings in an indefinitely frictionless fluid could be originated through such processes as we find setting up vortices in the imperfect fluids about us; on a perfect fluid such processes would have no hold. Now, questions of theoretical simplicity and definiteness apart, there is no gainsaying the fact that there is no experimental need for assuming this ether-matter to be a perfect fluid. No balance is delicate beyond six decimal places, and our longest astronomical records are but ephemeral in comparison with cosmical ages. Approximate constancy is thus all that our experience requires, and this we have by regarding the hypothetical fluid of the vortex atoms as indefinitely perfect; and have not, if we regard it as absolutely so.[1] Moreover, on the former alternative, we should be free to allow the possibility of ponderable matter coming to be here and ceasing to be there; the average amount in existence at once, either remaining stationary or else slowly altering, as is the case with the population of the globe, for example. Also we could entertain such a supposition without either flying in the face of any truth there is in what Mr. Spencer calls "the experimentally-established induction" that Matter is indestructible, or deserving his taunt of "not thinking at all, but merely pseudo-thinking."

This hydro-kinetic theory of matter as a mode of motion and not a substance, is specially wholesome and instructive, if we compare it with the modern theory of heat as a mode of motion, that has replaced the older theory of caloric as a substance. We cannot conceive substance to be either produced or destroyed, Mr. Spencer will tell us. True and trite, we must allow. When therefore it was found that heat and mechanical work were mutually transformable, there was an end of the theory that heat was a substance. It is now possible to produce vortex rings, to show that their behaviour in many respects approximates strikingly to the behaviour of material particles, and that this approximation would be greater if the fluids at our disposal were less unlike the continuous and frictionless fluid supposed to fill all space. Thus, though man may never be able to make or unmake a material particle, Lord Kelvin's ingenious speculations may at least pre-

[1] Cf. *Unseen Universe*, second edition, p. 118.

dispose us to believe in the thoroughly phenomenal character of all measurable masses, and, believing this, we are under no temptation to render absolute that relative constancy of such masses which our experience so far has disclosed.

How utterly unscientific it is to apply this principle of the conservation of mass to the entire universe is evident again when we reflect that it involves the further assertion that the universe is a finite system. Some recent writers on arithmetic talk of numbers that are at once infinite and complete, transfinite numbers as they are called. But it is obvious that there can be no scientific warrant for affirming such definite infinity of the universe, and there is certainly no empirical justification for affirming definite limits. No doubt what we see is limited; but to contend that we see no more, simply because there is no more to see, would be more illogical than it is to maintain that the bulk that may be beyond us *must* resemble the sample that we know. What we see is limited indeed in the sense of being finite, but it is not limited in the sense of being either constant or complete.

But now if the physicist were to ask the mathematician to devise for him a descriptive apparatus adapted to the movements of a material system in which the mass-values varied, the mathematician's first question would be: How do they vary? The physicist could not say. Innumerable forms of regular increase or decrease or of periodic alternation of the two are possible. Over against this bewildering variety the one definite supposition of constancy, in itself the simplest, is borne out by the very small fraction of the world that we can imperfectly measure. This seems to me how the case stands; and if it is, then it becomes plain that abstract dynamics affords as little ground for absolute statements about the magnitude or constancy of mass as for such statements concerning space or time. There are writers, however, who do not hesitate to rest this doctrine of the conservation of mass on that of the conservation of energy. But as this only means that in their opinion the latter doctrine cannot be true if the mass of the universe is not constant, such a plea is worthless unless there are independent reasons for maintaining that the energy of the universe is constant; and would not necessarily be true even then. The discussion of this important subject it will be best to defer till we have dealt with the application of abstract dynamics to the phenomena of molecular physics. To this I will ask your attention in the next lecture.

LECTURE IV

MOLECULAR MECHANICS: ITS INDIRECTNESS

Distinction of mass and molecule. The molecule not a 'minute body.'

The advance from abstract mechanics to molecular physics: Mechanics historically a usurper.

Molecular mechanics is (a) *indirect and* (b) *ideal.*

(a *i.*) *The evidence for* molecules *examined. Clerk Maxwell's theory of 'manufactured articles.' Clifford's criticisms. Further criticisms. Maxwell's theistic bias. The status of the molecule hypothetical. Statistical physics commented upon.*

(a *ii.*) *Evolution applied to the molecule. The mechanical theory bound, if possible, to resolve it into something simpler: the prime-atom.*

(a *iii.*) *The* ether — *one or more. Lord Kelvin sure of it, but chiefly because the mechanical theory cannot get on otherwise. New ethers invented to meet new mechanical problems. Signs of a reaction. Professors Drude and K. Pearson quoted. Hypothetical mechanisms and illustrative mechanisms distinct, but apt to get confused. Masterful analogies dangerous: is nothing intelligible but what is mechanical?*

THERE is no obvious similarity between the swinging of a pendulum or the motion of colliding billiard balls and the light and warmth of a glowing coal or of the sun. Still, as we have seen, Newton entertained the hope that both kinds of process might be described by means of the same mechanical principles. This hope we find has become an axiom for modern science; and the special conceptions involved and the peculiar methods employed in thus applying mechanical principles to molecular physics are what we must endeavour to examine to-day.

The distinction of mole and molecule, of large mass and small mass, is clearly not in itself a distinction of kind. It is due in the first instance to a psychological fact entirely external and irrelevant to the pure science of mechanics, to the fact, I mean, that we cannot perceive bodies of less than a certain size, changes of position of less than a certain extent, intervals of time of less than a certain duration, and so on. Still, however irrelevant to the mathematician, the fact of such *minima sensibilia* necessarily entails important differences of method upon the physicist, when he essays to apply mechanical principles to systems whose parts and motions are no longer directly discernible. The use of artificial means of magnification convinces us of what was already *a priori* probable, viz.: that the limits imposed by our senses are merely accidental limits without any objective significance. Consider in this connexion two statements that we often hear: the one that a given mole or molecule is divisible without limit into ever smaller particles; the other that such given mass or molecule consists of a finite number of absolutely indiscerptible particles called ultimate atoms. It is the latter far more than the former of these propositions that is logically open to suspicion. For the latter is an absolute statement, and since it is an absolute statement that cannot claim to be a necessity of thought, it is one that seems clearly incapable of proof. But to propositions of the former type, propositions, that is to say, asserting or implying the existence of bodies of indefinitely small dimensions and perhaps of indefinitely great complexity, we can have at any rate no *a priori* objection.

The molecules of modern physics and the so-called chemical atoms, however, are not bodies in this sense, and it is difficult to imagine that much would be gained by the assumption of their existence, if they were. This may sound paradoxical; I will try to explain. There is a passage in Laplace's *Exposition du systeme du monde*, one that has excited some discussion recently, which will serve admirably to illustrate what I mean, for it supposes an extreme case. Referring to the law of actions varying inversely as the square of the distance, as the law that holds for all forces and emanations that set out from a centre, he remarks: "Thus this law, answering exactly to all the phenomena, is to be regarded, both on account of its simplicity and its generality, as a rigorous law. One of its remarkable properties is that if the dimensions of all the bodies of the universe, their mutual distances and their velocities, were to increase or diminish proportionally, they would describe curves entirely similar to those they describe now; so that the universe thus continuously reduced down to the smallest space imaginable would present always the same appearances to observers."[1] If then we can have the universe on any scale, we might—if it is finite, as Laplace inclined to think it—have it complete within the head of a pin; and ought therefore to feel no surprise at physicists who, on the one hand, compare 'a compound atom,' as Jevons does, to a stellar system, each star a minor system in itself; "or who, on the other, talk of

[1] *o.c.*, bk. v, chap. v *fin.*, *Œuvres complètes*, 1893, vol. vi, p. 471.

Jupiter and his satellites as a planetary molecule."[1] But if a molecule were a constellation on a vastly smaller scale, then the phenomena of light, heat, magnetism, and the like, to explain which the molecular constitution of bodies has been assumed, would reappear in the molecule, and again in the molecule of the molecule, and so on indefinitely. On such lines then no logical advance could be made. There may be molecules or atoms of many orders, but, effectively to replace physical properties by mechanical processes, the molecule of any order must be divested of whatever property its motions are to explain or describe. Thus the molecules whose motions on the kinetic theory of heat answer to that state of a body which we call its temperature are not themselves credited with heat. Again, magnetism is not explained by resolving the smallest steel particles in a magnet severally into magnets, but by an imponderable fluid circulating round the particle, and so on.

Let us now attempt to characterise in a general way the application of abstract mechanics to molecular physics. We start with bodies of sensible dimensions. The dynamical transactions between such bodies can be directly observed and described, such description requiring no conceptions beyond those of mass, force, space, and time, except of course, number, which measurement involves. In confining itself to these conceptions, molar physics employs methods that are invariably abstract. Those important qualities possessed by every body in its own specific fashion, the differences of which remain for our perception as unique and irresolvable as are the sensations of

[1] Cf. Stallo, *Concepts of Modern Physics*, p. 122.

our several senses, — all these it simply ignores. They receive their proximate scientific handling in the various branches of experimental physics, *e.g.*, chemistry. Here numerous empirical laws are ascertained that do not in general overstep the qualitative barriers just mentioned. These comparatively restricted generalisations, obtained from experiments on light, heat, electricity, chemical composition and decomposition, and the like, are the material to which the theoretical physicist applies his mechanical scheme of molecules, molecular motions, and molecular forces. No doubt by this time the mathematical physicist himself undertakes or initiates experiments for the purpose of verifying or advancing his molecular constructions. But this in no way affects the fact that molecular physics can never come to close quarters with its molecules as molar physics can with the sensible masses and motions, from which the principles of the mechanical theory were first of all deduced.

To put the case in another way. Molar physics or mechanics was historically but one branch of general physics coördinate with those other experimental branches called Optics, Acoustics, Thermotics, etc. So matters stood in Newton's time, when he completed the main outlines of that mathematical edifice, now known as abstract dynamics, or, as he called it, 'rational mechanics.' Molecular physics is then, historically regarded, nothing but the endeavour to include the less perfect branches of physics within the domain of the most perfect — an endeavour that Newton himself, as we have seen, fully approved. The discovery that the stresses between electrified or magnetised bodies also varied inversely as the square of the distance between them, as do the stresses between gravitating masses, led to a wider use of the conception of centres of attraction or repulsion. Thus the mechanism which Newton found exemplified in the case of the heavenly bodies came to be regarded as a sort of type or paradigm. It would apply, as we have seen Laplace pointing out, on any scale, however great or small. So we come by the general hypothesis of molecular physics: that all physical phenomena—however complete, however ultimate, however numerous, their qualitative diversities may be, and remain, for our perception—can still be shewn to correspond to, and to be summed up by, purely dynamical equations, such equations describing the configurations and motions of a system of masses called molecules from their minuteness (according to the *Homo Mensura* standard). In other words, the hypothesis of molecular physics is that all the qualitative variety of the external world can be resolved into quantitative relations of time, space, and mass, that is of mass and motion.

This general characterisation of molecular physics we may now resume under two heads, each of which it will repay us to discuss somewhat further. First of all, the descriptions of molar physics may be called direct, whereas those of molecular physics are always indirect, the indirectness being often, if I may say so, of many removes from directness. Secondly, the descriptions of molar physics are abstract: *one* property of bodies, that of massiveness, of which we can have sensible evidence, is taken; the *remaining* properties are simply left out of account. But the descriptions of molecular physics taken together are not in this sense abstract. They leave no properties out of account; on the contrary, they transform everything qualitative into quantitative equivalents. It was to this point that I referred at the outset of this discussion (in the second lecture) in calling the methods of molecular physics ideal.[1] I should be glad of some less ambiguous term, but can only hope that at the end of our discussion its meaning may be clearer.

To begin with the indirectnesses. Nobody has ever seen or felt, and if the physicists are to be trusted, no instruments of magnification are possible by which in the future any one can be helped to see or feel, an individual molecule. This, of course, would be a matter of no importance if the molecule were merely regarded as a mass-element in some homogeneous mass of sensible volume. But the atoms and molecules of modern science, if they have any real existence at all, are distinct individuals; at all events, they have more title to be so described than either the earth or the sun, which we commonly regard as individual objects. For the earth or sun are after all but aggregate masses, constantly receiving additions — as in the meteoric showers that feed the sun; and probably — in the case of the earth and many smaller bodies, at least — constantly scattering part of their mass into space, as the moon, for example, is supposed to have diffused away its free gases and vapours. Not so the atoms and molecules of the chemist. The progress of stellar spectroscopy and

[1] Cf. above, p. 51.

of chemical physics, we are told, shuts us up to the view that the whole universe apart from the ether or ethers — of which more presently — consists entirely of varying arrangements of incalculable numbers of some seventy different elements, the individuals of each kind being absolutely identical in their properties, and all alike entirely beyond the reach of change or decay. Philosophic speculations of this sort are, of course, no novelty; but when we are asked to accept such statements as scientific truth and verity on evidence that *can* only be indirect, we may well be pardoned by 'those who know' if we look a little critically, even sceptically, at that evidence. But you may wish first of all to have the statement itself in some accredited form. Let me then quote two or three sentences from the *Collected Papers* of Clerk Maxwell (vol. ii, pp. 361 ff.): — "The same kind of molecule, say that of hydrogen, has the same set of periods of vibration, whether we procure the hydrogen from water, from coal, or from meteoric iron. . . . Whether in Sirius or in Arcturus [it] executes its vibrations in precisely the same time." "Though in the course of ages catastrophes have occurred, and may yet occur, in the heavens, though ancient systems may be dissolved and new systems evolved out of their ruins; the molecules out of which these systems are built — the foundation stones of the material universe — remain unbroken and unworn." Elsewhere Maxwell proceeds to make inferences concerning the supernatural from this position. "None of the processes of Nature," he says, "since the time when Nature began, have produced the slightest difference in the properties of any molecule. We are therefore unable to ascribe either the existence of the molecules or the identity of their properties to the operation of any of the causes which we call natural. On the other hand, the exact equality of each molecule to all others of the same kind gives it, as Sir John Herschel has well said, the essential character of a manufactured article, and precludes the idea of its being eternal and self-existent." This argument would be open to question even if it were certain that the molecules of any given element are *exactly alike*. To many it would seem more reasonable in such case to side with Democritus and regard what within the whole range of actual or possible experience is absolutely permanent and without the shadow of a change as realising all that we can understand by 'self-subsistent and eternal.' Moreover, the disparity between the conception of creation and the conception of manufactured goods is so complete as to make all attempts at analogy futile.

But to return to our immediate question: Of what nature is the evidence, on which molecules of hydrogen, oxygen, or any supposed element are pronounced to be respectively, each to each, exactly alike, the same through all vicissitudes and everlasting as time itself. As to the exact likeness — let me once more remark that it is impossible to deal directly with the individual molecules; and, even if it were, no measurements and no physical comparisons are exact. But the measurements of molecules, besides being indirect, are all made in bulk. What is really measured is the combined effect of millions, or it may be billions, of molecules.

So that, even supposing disturbing causes to be entirely excluded, the resulting measurement is true only of the average molecule and leaves the range of the individual deviations at best but partially determined. The most delicate test so far available, that of the spectroscope, seems always to be beset by at least one disturbing factor. On this method the qualitative identity of the molecules of a given element in the gaseous state is inferred from their light-note. But every one who has heard the sound-note of the whistle of a train in motion must have observed that this note sounds higher so long as the train is approaching, and lower as soon as it has passed and begun to recede. To get the light-note true, the molecules should be observed free from their translatory motions towards and away from the observer. The variations thus produced can only be set down entirely to the account of the translatory motions after independent proof has been adduced of the absolute likeness of the molecules. Meanwhile it has to be shared between the two. But since Maxwell wrote the passages I have quoted, it has been shewn that the spectra of several elements vary with the temperature and the pressure to which the gas is exposed; and when a gas approaches the liquid condition these changes appear to be greater still. What various degrees of aggregation there may be in the liquid or solid state, and how far the individuality of the molecule disappears in such aggregation — these are problems for which there appears at present no definite solution.[1]

Graham's familiar method of dialysis, or atom-sifting,

[1] Cf. Ostwald, *Outlines*, pp. 189, f.

is also appealed to by Maxwell to establish the perfect identity of the molecules of the same kind of matter. Graham found, it will be remembered, that light gases pass through a porous septum more rapidly than heavier ones. Maxwell is referring to this method when at the close of his book on *Heat* he says: "If of the molecules of some substance such as hydrogen, some were of sensibly greater mass than others . . . in this way we should be able to produce two kinds of hydrogen, one of which would be somewhat denser than the other. As this cannot be done, we must admit that the equality which we assert to exist between the molecules of hydrogen applies to each individual molecule, and not merely to the average of groups of millions of molecules."[1] But there is a world of difference between saying of a million molecules that the mass of no one of them is '*sensibly greater*' than that of the rest, and saying that the masses of all are absolutely equal.

I cannot help thinking that Clifford reasons far more soundly than Maxwell in dealing with this same method of dialysis. "If we put any single gas into a vessel," he says, "and we filter it through a septum of black-lead into another vessel, we find no difference between the gas on one side of the wall and the gas on the other side. That is to say, if there is any difference, it is too small to be perceived by our present means of observation. It is upon that sort of evidence that the statement rests that the molecules of a given gas are all very nearly of the same weight. Why do I say *very nearly?* Because evidence of that sort can never prove

[1] *Heat*, p. 339.

that they are exactly of the same weight. The means of measurement we have may be exceedingly correct, but a certain limit must always be allowed for deviation; and if the deviations of molecules of oxygen from a certain standard of weight were very small, and restricted within certain limits, it would be quite possible for our experiments to give us the results which they do now. Suppose, for example, the variation in the size of the oxygen atoms were as great as that in the weight of different men, then it would be very difficult indeed to tell by such a process of sifting what that difference was, or, in fact, to establish that it existed at all. But, on the other hand, if we suppose the forces which originally caused all those molecules to be so nearly alike as they are to be constantly acting and setting the thing right as soon as by any sort of experiment we set it wrong, then the small oxygen atoms on one side would be made up to their right size and it would be impossible to test the difference by any experiment which was not quicker than the process by which they were made right again."[1] Had Clifford been writing now he might have illustrated this last point by a reference to Mr. Galton's principle of reversion towards the mean, in accordance with which the children of giants, for example, tend to be of less stature, and the children of dwarfs to be of greater stature, than their parents.[2]

[1] *Lectures and Essays*, vol. i, p. 207.

[2] It is well known that some chemists agree with Sir William Crookes in thinking that "probably our atomic weights merely represent a mean value around which the actual atomic weights of these atoms vary within certain narrow limits," reminding us of Newton's 'old worn particles,'

But Maxwell felt himself "debarred from imagining any cause of equalisation on account of the immutability of each individual molecule"—this being the second article of his molecular creed, as that of exact likeness was the first. There is, I fear, something circular in Maxwell's arguments for these two positions. On the one hand the ingenerability and immutability seem to be used in proof of the qualitative and quantitative identity; although, on the other, this very identity had served as an argument for that everlasting constancy which in turn it now helps to prove. Nay, his argument seems even weaker than that, for he takes for granted that the persistence which he asserts for his normal molecules would belong also to abnormal ones, if any such there were. And so, assuming the exact equality of all the individual molecules of hydrogen, etc., within the range of our experience, he asks where can the eliminated molecules have gone to? He then proceeds: "The time required to eliminate from the whole of the visible universe every molecule whose mass differs from that of some of our so-called elements, by processes similar to Graham's method of dialysis, which is the only method we can conceive of at present, would exceed the utmost limits ever demanded by evolutionists as many times as these exceed the period of vibration of a molecule." But surely it is quite gratuitous.

save that the result is not supposed to be due to wear and tear. Besides referring to Sir William Crookes's researches into the fractionation of yttrium—one more instance, and a splendid one, of the saying that genius is patience—I may mention the experiments on the homogeneity of helium just published by Messrs. Ramsay and Collie. See *Nature*, 1896, vol. liv, p. 408.

to assume that they could only disappear by being sifted out on some chaotic dust-heap beyond the fixed stars, a sort of limbo for manufactured articles spoilt in the making.

And this remark suggests a more searching question: What, precisely, is it of which this immutable individuality is affirmed? Is it of a form or is it of a substance? The biologist can tell us of species that have persisted unchanged from times so long anterior to ours that the hoariest mountain ranges appear by comparison to have sprung up but yesterday. But here it is only the *form* that endures, the particular individuals being quite transitory. A lake dries up and its tiny inhabitants perish; after a longer or shorter interval the water returns and the old living forms reappear. But the biologist does not follow the analogy of the chemist, and pronounce these to be necessarily the earlier individuals emerging from some quasi-chemical condition in which their characteristic properties have been suspended or masked. Now physical astronomers find that the spectra of certain of the whiter, and presumably hotter, stars yield indications of no element save hydrogen; also that as stars approximate to a red colour, and so have presumably a lower temperature, they furnish more varied and complex spectra, indicating the presence of many other elements besides hydrogen. The simplest supposition we can make—and it is one actually made—is that in the earlier stages of stellar evolution, of which we thus get peeps, the various chemical elements come successively into being, as do various forms of vegetable and animal life in the later stages of the same vast process.[1] But what becomes of the molecule as an article manufactured before natural processes began? The best that can be said is, not that the individual article is a fabric of timeless origin, but only that its form or pattern is thus (relatively) immutable and ingenerable. It is still possible, however, to reinstate some persisting individual by falling back on primal atoms or elements of a higher order. And phenomena daily observed by the chemist at once suggest this step. As ordinary chemical compounds can be decomposed at high temperatures, it is probable that our so-called elements may be 'split up' into elements of a new order by temperatures greatly in excess of any that we can command. Those who think fit may regard this higher order of element as furnishing "the foundation stones of the material universe" and remaining—though the firmament be dissolved and renewed again—"in the precise condition in which they first began to exist." But such an opinion can no longer be entertained of the molecules 'built up' of these stones —molecules that processes now going on seem to make and unmake, as the chemist makes further compounds out of them, which he can afterwards decompose again. Maxwell was evidently prepared for this alternative. In the closing paragraph of his *Theory of Heat*, he asks, "But if we suppose the molecules to be made at all, *or if we suppose them to consist of some thing previously made*, why should we expect any irregularity to exist among them?"

But surely it is far from indifferent which of these

[1] Cf. Sir W. Crookes's brilliant Address to the Chemical Section of the British Association, 1886, *Nature*, vol. xxxiv, pp. 423 ff.

alternatives we adopt when inquiring what amount of "irregularity" we may expect among the molecules of any given chemical stuff. If the molecules of oxygen, hydrogen, etc., are themselves primeval and immutable individuals, they are like nothing else that we know, and we can have no scientific grounds for *expecting* anything about them one way or other. But if they are compounds that are put together and again 'split up' in the course of nature, then, in the absence of certain knowledge to the contrary, we *may* expect among their forms any of the regularities or irregularities that we find elsewhere among dissoluble products. In particular we might expect, for example, that certain of these forms, like some of the chemical compounds that we know as such, would prove very unstable, and so disappear almost as soon as they arose; others again, like certain refractory minerals long regarded as elements, might persist indefinitely. The striking analogy between the grouping of chemical elements, when ranged as in the periodic laws of Meyer and Mendelejeff, and the grouping of biological forms, might tempt us to entertain the hypothesis, *mutatis mutandis*, of some sort of chemical evolution. But absolute qualitative identity, for which Herschel and Maxwell contended, would be almost as incompatible with such an hypothesis as absolute immutability. Both these absolute ideas would be alien to the notion of continuous transmutability or of connecting forms.

Digressing for a moment, let me remark that both these ideas, there can be little doubt, are far more due to theological zeal than to the bare logic of the facts. In the fine conclusion of his text-book on *Heat*, after asking, "Why should we expect any irregularity to exist among them,"—the molecules, *i.e.* of the same kind of matter,—Maxwell continues: "Why should we not rather look for some indication of that spirit of order, our scientific confidence in which is never shaken . . . and of which our moral estimation is shown in all our attempts to think and speak the truth, and to ascertain the exact principles of distributive justice?"[1] But why so confidently assume, we might reply, that a rigid and monotonous uniformity is the only, or the highest, indication of the spirit of order, the order of an ever-living Spirit above all? How is it then that we depreciate machine-made articles and prefer those in which the artistic impulse or the fitness of the individual case is free to shape and to control what is literally manufactured, hand-made? The work of an engine-fitter is greatly facilitated by the use of Whitworth bolts, tubing of regulation sizes, and the like, but surely it is trivial to frame teleological arguments concerning the universe from the standpoint of a millwright. So the existence of a limited number of absolute constants in nature might bring the universe within the compass of the Laplacean calculator. But, dangerous as teleological arguments in general may be, we may at least safely say the world was not designed to make science easy. Struggling men and women, like the soldier on the march when his machine-made shoe pinches, might reasonably complain if science should succeed in persuading them that Nature's doles and Nature's dealings from first to last are ruthlessly and rigidly mechanical. To

[1] *Heat*, p. 342.

call the verses of a poet, the politics of a statesman, or the awards of a judge *mechanical*, implies, as Lotze has pointed out, marked disparagement: although it implies, too, precisely those characteristics — exactness and invariability — in which Maxwell would have us see a token of the Divine.

But, returning to our facts and avoiding altogether any question as to why we should expect this or why we should expect that, for such questions lie beyond the legitimate pale of science, let us gather up what we find. Chemical molecules are not presented realities: in other words, a molecule — say of oxygen — is not a small body which is known to exist as an individual of a definite species, distinct, say, from a molecule of nitrogen, an individual of another definite species of small body. Individual chemical molecules are not known, as rubies or palms are known, *i.e.* as instances of species and distinct from diamonds or cedars, instances of other species. The chemical molecule is a hypothetical conception. Such things *may* exist or the hypothesis would not be legitimate. Whether they actually exist or not, they, at any rate, serve, like certain legal and commercial fictions, to facilitate the business of scientific description. If they exist, then facts show that the molecules of a given species are very nearly alike; the said facts admitting of interpretation according to statistical methods. As in other cases admitting of statistical treatment, so here the physicist is free to regard all molecules of a class as exactly like his mean or average molecule. But he is not entitled to let this abstract simplification harden into concrete fact. Perhaps it may be thought that such rigorism is pedantic. So far as any particular physical inquiry is concerned it may be, but I am very doubtful even of this. At all events, if such unwarrantable concreting of abstracts is to lead logically to a mechanical theory of the universe, we do well to take note of it.

To make the bearing of this remark clearer, let us turn our attention for a moment to the very parallel case of economic theory and the interpretation of industrial and social statistics. The science of so-called pure or deductive economics has much in common with physics, that is to say, it sets out from definitions and axioms and seeks to describe economic facts by means of mathematical equations. The 'economic man' as conceived by Ricardo, a 'market' as defined by Cournot, James Mill's 'doses of capital,' the 'margin of cultivation,' or Jevons's 'supply and demand curves,' are not things we expect to meet with in real life. They are abstractions that summarise experience, not concrete realities directly experienced. Englishmen about to marry are not observed to be exclusively interested in women their juniors by 2.05 years, though according to the tables this is the difference of age between the Englishman and his wife. But, again, the Englishman or the Frenchman, or the civilised man or the savage, is a concept, not a reality. Yet a science of anthropology is possible in which different races of men and different stages of human development are compared by the help of mean values obtained by dealing with nations and societies *en bloc*. And perhaps "in this way," as Lotze has said, "we may easily imagine how all kinds of formulæ may be arrived at, expressive of the acceleration and breadth and depth and colouring of the current of historical progress, formulæ which, if applied to particulars, would be found to be utterly inexact, but which can yet claim to express the true law of history as freed from disturbing individual influences." It was precisely this misapplication to particulars that led Buckle to say that in a given state of society a certain number of persons *must* put an end to their own lives. Now, if, when both the varying particulars and the statistical constants are alike well known, it is possible for a reasonable man to fall into the error of converting the one into an iron necessity which rules over the other, no wonder this should be the prevalent attitude in departments of knowledge where particulars are beyond our ken. I contend then that the most the physicist is entitled to assert is, that, if there are molecules, the mass of the mean oxygen 'atom' is sixteen, that of the mean hydrogen 'atom' being taken as unity; and so on for the rest of his table of masses. He is not entitled to say that if there are molecules the mass of every oxygen atom is precisely sixteen times the mass of any hydrogen atom. Try to picture to yourselves the sort of science of man and of society that would be formulated by an intelligence whose data were confined to anthropometrical and other statistical results and who treated his data in the customary physical fashion. You will conclude, I think, that his human beings or *homunculi* would come out surprisingly like Herschel's molecules as 'manufactured articles,' and that his theory of society would have more than a superficial resemblance to the kinetic theory of gases.

Finally, as the facts do not justify the assertion of exact likeness among molecules, neither do they afford ground for the assertion that individual molecules are immutable and incorruptible. Once this is clear, then molecules, if there are such things, come within the range of the great conception of evolution and facts pointing in this direction are known already and are steadily accumulating. As Huxley well says: "The idea that atoms are absolutely ingenerable and immutable 'manufactured articles' stands on the same sort of foundation as the idea that biological species are 'manufactured articles' stood thirty years ago; and the supposed constancy of the elementary atoms, during the enormous lapse of time measured by the existence of our universe, is of no more weight against the possibility of change in them . . . than the constancy of species in Egypt since the days of Rameses or of Cheops is evidence of their immutability during all past epochs of the earth's history. It seems safe to prophesy that the hypothesis of the evolution of the elements from a primitive matter will, in future, play no less a part in the history of science than the atomic hypothesis, which, to begin with, had no greater, if so great, an empirical foundation."[1] We may, I think, go even farther. Somehow or other the qualitative diversity of the chemical elements must admit of description by means of quantitative relations of mass-points, configurations, and movements—*if the mechanical theory is to make good its claims.* Indeed,

[1] *Collected Essays*, vol. i, pp. 79 f.

the unceasing efforts of chemists and physicists in this direction can be regarded as an emphatic admission that they have laid this charge upon themselves. Moreover, in what is called the New Chemistry or General Chemistry — take Ostwald's well-known *Outlines* as an example — we see how much they have already accomplished; and also, I will add, how very much more still remains to be done.

But let us turn now to another order of facts. If the molecules concerned in chemical reactions and in the kinetic theory of gases are beyond sensible reach, the forms of matter immediately concerned in the phenomena of radiation, electricity, and magnetism are more remote still. It is in connexion with these that the ether or ethers come upon the scene. I say ethers because it is by no means certain that one will suffice. "It is only when we remember," says Maxwell, "the extensive and mischievous influence on science which hypotheses about ethers used formerly to exercise, that we can appreciate the horror of ethers which sober-minded men had during the eighteenth century, and which, probably as a sort of hereditary prejudice, descended even to the late J. S. Mill." Time seems to have brought its revenge, for nowadays the ether is regarded as preëminently real. Thus, in a lecture given about ten years ago and recently published, our foremost physicist said to his hearers: "You can imagine particles of something, the thing whose motion constitutes light. This thing we call the luminiferous ether. *That is the only substance we are confident of in dynamics.* One thing we are sure of, and that is *the reality and substantiality of the luminiferous ether.*"[1] Yet in spite of this confidence of Lord Kelvin's I cannot help thinking that a jury of logicians would side with Mill. But possibly some of you may be disposed to ask, What has the question as to the real or hypothetical nature of the luminiferous ether to do with the mechanical theory of the universe? Simply that unless a material medium for its propagation is either found or assumed, the phenomena of light cannot be mechanically described. And the remark applies equally to other forms of radiation as well as to electricity and magnetism. If not themselves massive, these phenomena must depend on the configuration or motions of something that is massive, or it is obviously impossible to describe them in the mechanical terms at present in vogue. That need entail no detriment to the special physical sciences concerned with their description and measurement by means of a more concrete and qualitative terminology; and, indeed, some able physicists prefer to leave the question of a medium entirely aside.[2] But to do this so far puts a stop to the resolution of all physical changes into mechanical processes. We shall all perhaps allow a reasonable presumption in favour of any theory that will unify the variety of physical facts. Perhaps some of us feel that physicists have too hastily assumed that, unless these facts have a common mechanical foundation, they can have no intelligible connexion at all. Even if the mechanical theory turn out to be true in fact, there is no *a priori* neces-

[1] Lord Kelvin, *Popular Lectures and Addresses*, vol. i, p. 310.

[2] F. E. Neumann for example. Cf. Volkmann, *Theorie des Lichts*, p. 4.

sity about it. Yet covertly or overtly some such necessity is assumed; and it is mainly on the basis of this postulate that the ether is raised from the subsidiary position of a descriptive hypothesis to the rank of a thing having "reality and substantiality." Grant, first, that the world must be intelligible; grant, secondly, that to be intelligible it must be mechanical; and then grant that to be mechanical there must be an ether or ethers whose motions constitute light, electromagnetism, etc., grant all this and then — spite of the absence of direct evidence — we might say the existence of ether is indirectly proved. But the first two steps in this argument, it will be observed, are philosophical and the second very disputable philosophy. Science, however, has no right to build on philosophical premisses, and is forward, as we have seen, to disown, with much needless blasphemy, all such *a priori* methods. Leave aside then any presuppositions of this kind, and the ether remains but a mechanical hypothesis; its perceptual reality, if proved at all, can only be proved by some crucial experiment or by cumulative experimental evidence. No doubt its value as a descriptive hypothesis has been greatly enhanced since Mill's time — notably by the labours of Maxwell and Hertz. But as to the worth of their results I suppose Poincaré's remark upon it is not too cautious: "There still remains much to be done; the identity of light and electricity is from to-day something more than a seducing hypothesis; it is a probable truth, but it is not yet a proved truth."[1]

[1] *Nature*, 1894, vol. l, p. 11.

But though the conception of an all-pervading ether has gained in scientific importance since Mill's controversy with Whewell, it has also been repeatedly modified, I might even say transformed. At one time or other it has been regarded as a gas, as an elastic solid of small density but high rigidity, as a 'quasi-solid' constituted by turbulent motion in an incompressible inviscid fluid — with two or three sub-varieties of this hydrokinetic type. And when a new ether is invented the problem is to ascertain how many of the special laws of radiation or electricity can be mechanically deduced from it. In no case has this demand been adequately met; hence the attempts, continually renewed, to devise more satisfactory ethers. Surely if the ether were a *definite* thing, the reality of which was an established fact, it would be impossible to take these liberties with it. On the other hand, is it not certain that if, conceivably, some non-mechanical hypothesis were to afford a simpler and more complete unification of optical and electrical phenomena, there would be an end of luminiferous and electric ethers, just as there was an end of phlogiston in the days of Priestley and Lavoisier, and as there has been an end of caloric and electrical fluids in our own? By a non-mechanical hypothesis, I mean here one in which some or all of the Newtonian laws are denied or modified.[1] I should hardly have ventured even to suggest such a thing on my own responsibility. But I observe that several physicists in the present unsettled state of the science are prepared to entertain such heresies. I

[1] Perhaps such a restriction is in itself unwarranted, but it serves my purpose here.

will quote two. Professor Drude, on succeeding to a new chair at Leipzig, devoted his inaugural lecture to the *Theory of Physics*. Referring to the characteristic difference between what we call matter and what we call ether, viz.: that the former consists of smallest inhomogeneities, — a finely grained structure, as we say in English, — while the latter is thoroughly homogeneous, he continues: "The physics of matter must then appear the more complicated compared with the physics of the ether. Is not that an indication that no simplification can result if we attempt to describe the physics of the ether formally in the same manner as the physics of matter, that is to say, by means of mechanical equations?"[1] Again, Professor Karl Pearson, in his *Grammar of Science*, referring to the Newtonian laws, asks: "Ought we to assert that these laws hold in their entirety for all the scale from particle to ether-element? Or will it be more advantageous to postulate that mechanism in whole or part flows from the ascending complexity of our structure, that the ether-element is largely the source of mechanism, but is not completely mechanical in the sense of obeying the laws of motion as given in dynamical text-books?" And in another passage: "The object of science is to describe in the fewest words the widest range of phenomena, and it is quite possible that a conception of the ether may one day be formed in which the mechanism of gross 'matter' itself may, to a great extent, be resumed. Indeed, it is on these points of the constitution of the ether, and the structure of the prime atom, that physical theory is at present chiefly at fault.

[1] *Die Theorie in der Physik*, 1895, p. 13.

There is plenty of opportunity for careful experiments to define more narrowly the perceptual facts we want to describe scientifically; but there is still more need for a brilliant use of the scientific imagination. There are greater conceptions yet to be formed than the law of gravitation or the evolution of species by natural selection. It is not problems that are wanting, but the inspiration to solve them; and those who shall unravel them will stand the compeers of Newton and Darwin."[1]

The remarks and queries just quoted apply to the electric and luminiferous medium or media, though the medium the writers have also in view is doubtless what has been called "the primordial medium"; such, *e.g.*, as the perfect fluid of Lord Kelvin's vortex-atoms, from which ultimate ether the proximate ether of light and electricity is supposed to be formed. At this primordial and absolutely homogeneous fluid the physical theorist is content at last to stop; and for this at present no confident claim is advanced to "reality and substantiality." Will the physicists of fifty years hence remain as modest — should the hypothesis, as seems likely, hold its ground so long?

So much then must serve to illustrate what I called the indirectness of molecular physics. Under this head we have noted a tendency to treat statistical means and hypothetical mechanism as concrete realities. And here it seems needful to make a distinction or we may be charged with unfairness — a distinction, I mean, between hypothetical mechanisms and illustrative mechanisms employed solely for expository purposes. To the latter

[1] *Grammar of Science*, pp. 339–369.

class, for example, belong unquestionably the "idle wheels" of Maxwell's electro-magnetic theory and again Lord Kelvin's gyrostatic cells. On the other hand his quasi-elastic ether, or his quasi-labile ether, seem to be meant as real and not as merely illustrative analogies. But it is to be feared that physicists of the school of Maxwell and Lord Kelvin, who — to use Boltzmann's description of them — "are particularly fond of the variegated garment of mechanical representation," are apt unconsciously to play fast and loose with the difference between fiction and fact, when elaborating their mechanical models. Analogy, as we know, is a good servant, but a bad master; for, when master, it does more to blind than it may previously have done to illuminate. Most of us, I suppose, have chanced to observe a bee buzzing up and down within the four sides of a window-pane, vainly endeavouring to escape by the only obvious way — the way most light comes; whereas by merely traversing the dark border of the window-frame it might at once reach the open casement. The history of science is full of instances of able men similarly thwarted by a too-prepossessing analogy. In his lectures at the Johns Hopkins University Lord Kelvin is reported to have said, "I never satisfy myself till I can make a mechanical model of a thing. If I can make a mechanical model I can understand it. As long as I cannot make *a mechanical model all the way through*, I cannot understand, and that is why I cannot get the electro-magnetic theory of light."[1] Now I should like respectfully to ask whether

[1] *Nature*, vol. xxxi, p. 603.

this is not possibly a case of unwarrantable submission to analogies. As before, I ask again: Why must mechanism "*all the way through*" be the one and only means of intelligibility? When we recollect the comparatively small range of the experiences within which mechanical laws are found to be verifiable abstractions, are we bound to assume that they are the only concrete realities at the very foundations of physical things? This question brings us to the second characteristic of molecular mechanics just now referred to—its ideal of matter. The consideration of this may perhaps give us further light, but must be deferred till the next lecture.

LECTURE V

MOLECULAR MECHANICS: IDEALS OF MATTER

(b) *The* ideal *of matter. The old atomism strictly mechanical but inadequate. Its conversion into one strictly dynamical by Boscovich and the French. The resolution of this in turn into the 'kinetic theory.'*

The nature of the primordial fluid examined: it is made up of negations, and is thus indeterminate: prima materia.

Relation of its mass to the 'quasi-mass' of the vortices: the latter becomes a complicated problem. The kinetic ideal in danger from 'metaphysical quagmires.' To avoid this impasse *it is proposed to make energy fundamental.*

Results of inquiry into mechanical theory thus far: *Relation of the three sciences, Analytical Mechanics, Molar Mechanics, Molecular Mechanics. The first stands completely aloof from concrete facts. The attempt to apply it to these without reserve leaves us with a scheme of motions and nothing to move.*

To molar mechanics belongs the rôle *of stripping off the physical characteristics of sensible bodies; to molecular mechanics, the* rôle *of transforming these characteristics into mechanisms, and the mechanisms into 'non-matter in motion.' The mechanical theory as a professed explanation of the world thus over-reaches itself.*

As mechanical science has advanced, its true character has become increasingly apparent—its objects are fictions of the understanding, and not conceivably presentable facts.

The kinetic ideal shows this best of all, for some of its upholders dream of 'replacing' dynamical laws by kinematical. The refutation the more striking because they imagine they are all the while getting nearer to 'what actually goes on.'

It is upon an uncritical prepossession of this kind that the mechanical theory has rested all along. Descriptive analogies have been regarded as actual facts; yet are nothing but the inevitable outcome of the endeavour to summarise phenomena in terms of motion. A moral drawn from the Pythagoreans.

But mechanical science has so far failed even to describe *facts in its own terms.*

WE have found physicists protesting with great vehemence against being saddled with any metaphysical conceptions of matter as a substance underlying phenomena. Yet there is only one of the three chief theories of matter that might readily clear itself of this stigma, and that is the old atomic theory of Democritus or Lucretius; but this, oddly enough, has always claimed to be a theory of substance. In point of fact it is the most phenomenal of all; for the hard atom, apart from its being *absolutely* hard, differs from sensible bodies only in respect of size and indivisibility. The collisions of such atoms again are essentially phenomenal, though actually beyond the limits of direct perception. Such collisions too are the very type of that plain, straightforward mechanical action, which alone Galileo, Newton, and Huygens — the founders of modern mechanics — were willing to recognise. You will remember the often-quoted letter of Newton to Bentley, in which he declared it to be "inconceivable that inanimate brute matter should . . . operate upon and affect other matter without mutual contact." This then is logically the one genuine and original mechanical theory. But absolute hardness is ideal and transcends experience, whereas for the physicist bodies are real and empirically given. I think we may say that whoever ventures to apply to any real thing such adjectives as 'absolute' or 'infinite' or 'perfect' or 'simple'—the terms being strictly used—has, however much he may dislike it, entrangled himself with metaphysics. Such at least has been the fate of the Lucretian atom, when defined as absolutely hard. Whether Lord Kelvin's perfect fluid fares any better, we can consider later. But let us first notice some of the antinomies besetting the older ideal atom.

Rigid bodies of sensible dimensions are described as respectively elastic or non-elastic, according as they do or do not resume their original shape after being strained. Absolute rigidity, however, absolutely excludes deformation, hence the hard atom can neither be elastic nor non-elastic. What then will happen when two such atoms collide? The problem is strictly indeterminate, so that—as has been said—as often as such an event occurs, the course of the world is at least as uncertain as an act of the purest free will could make it.[1] "Take a series of very inelastic bodies such as butter, lead, etc.," says P. du Bois-Reymond, "and then a series of very elastic bodies, such as india-rubber, ivory, etc. Of which of these two series is the absolutely hard the limit? Obviously of which we like, or of some mean between both."[2] If we decide to regard the atoms as non-elastic, then, when two collide, kinetic energy disappears without an equivalent amount of potential energy taking its place. If we prefer to regard them as elastic, their motions are instantaneously reversed, in other words, a finite momentum is produced in no time. If we combine the two, we

[1] Kroman, *Unsere Naturerkenntniss*, p. 315.

[2] *Ueber die Grundlagen der Erkenntniss in den exacten Wissenschaften*, p. 37.

combine these consequences; both of which contradict our fundamental axioms. The fact is that rigidity, whether accompanied by much or little elasticity, is not a property of mass as such, but a physical property of matter. But if a physical property, then rigidity has to be explained by dynamical transactions between masses, or the mechanical theory fails to redeem its pledge. In other words, it is not open to the physicist to explain—or, as is now said, to describe—rigidity and elasticity in terms of rigidity and elasticity. The retention or restitution of a given shape or configuration implies mechanical or dynamical relations between masses and has to be accounted for. So by inexorable logic the "many hard, impenetrable particles," which Newton was content to regard as "primitive," were resolved step by step into the mass-points or centres of force of Boscovich and the French analysts. But as contact action, *i.e.* action of the straightforward mechanical type, is impossible between mass points, it was replaced by action at a distance, sometimes attractive, sometimes repulsive, according as the distance or other circumstances might vary. The strictly mechanical theory became in fact strictly dynamical.

A word or two of historical explanation seems called for as to this opposition between two terms—I mean 'mechanical' and 'dynamical,' which are nowadays often regarded as synonymous. The term 'mechanical' however, seems appropriate only to motions produced by immediate displacement, as in machines, to contact action in other words. Newton, who—as we have seen—regarded action at a distance as "so great an absurdity, that I believe," he writes to Bentley, "no man who has in philosophical matters a competent faculty of thinking can ever fall into it," finding himself unable *mechanically to explain* the working of gravitation, contented himself meanwhile with *describing* the motions produced. But he began early and persisted long in the attempt to discover some medium and mode of operation such as would enable him to explain gravitation by contact, instead of assuming it to be a force "innate, inherent, and essential to matter." However his friend and contemporary, the youthful Roger Cotes, though anything but a fool, rushed in where the master feared to tread. In his preface to the *Principia*, Cotes definitely asserted the doctrine of direct action at a distance, and maintained that gravity is no more an occult property of matter than extension, mobility, or impenetrability; since it was, he held, as plainly indicated by experience as they were. "And when" — I here quote Maxwell — "the Newtonian philosophy gained ground in Europe, it was the opinion of Cotes rather than that of Newton that became most prevalent, till at last Boscovich propounded his theory that matter is a congeries of mathematical points, each endowed with the power of attracting or repelling the others according to fixed laws. In his world, matter is inextended, and contact is impossible. He did not forget, however, to endow his mathematical points with inertia."[1] Thus Newton's position was exactly inverted. The solid primitive particles of various sizes and figures, in which Newton inclined to believe, were rejected; and the inherent forces acting through a vacuum, which he

[1] *Scientific Papers*, vol. ii, p. 316.

disclaimed as absurd, were accepted as the reality to which all the physical properties of matter were due. This is what I meant by saying that his strictly mechanical theory was transformed into one strictly dynamical.

One step in this transformation seems, as I have said, logically inevitable, the reduction of finite molecules to infinitesimal mass-points. Not so the second — the attribution to such mass-points of intrinsic forces. We have seen already that in abstract mechanics this conception of *vires insitæ* or substantial forces is rigorously scouted. Force is there a purely relative conception, a name for the rate of change of momentum of one mass referred to the position of other masses in the same "field." Unless then Boscovich's metaphysical idea of forces inherent in a mass-point can be replaced by the mathematical idea of external forces acting at a point, molecular physics cannot be regarded as merely dynamical in the looser modern sense. Central forces when not used geometrically, as by Newton, *i.e.* merely to *describe* observed motions, but metaphysically, to explain *action at a distance*, are incompatible with modern mechanics. They become part of what Professor Tait calls a "very old but most pernicious heresy, of which much more than traces still exist even among physicists."[1] But it must certainly be allowed that the progress of physics has steadily discredited it. Faraday's experimental researches into electricity and magnetism, the resolution of heat into "a mode of motion," and many other lines of investigation tend to confirm the kinetic ideal of matter, which has been aptly described

[1] *Properties of Matter*, p. 6.

as the theory that matter is non-matter in motion — the non-matter, being of course, Lord Kelvin's ideal fluid.

It is this kinetic, or perhaps I should say hydrokinetic, ideal of Lord Kelvin and his school, that, so far as I can gather, is the ideal of matter prevalent in the present day among such physicists as venture to stir beyond their equations. Any one with a weakness for Hegelian dialectic might easily discover the famous triadic development of thought in the advance from what was in the main Newton's ideal of matter through the ideal of Boscovich to that of Faraday and later British physicists. There seems to have been complete opposition between Newton's conceptions as to what matter really was and the descriptive apparatus of central forces acting across empty space by which he simplified and extended the more cumbrous apparatus of Kepler. Boscovich's doctrine was thus the precise antithesis of Newton's, for he took Newton's descriptive apparatus for the reality, and discarded his solid, impenetrable particles as false. Boscovich's atoms were strictly mass-points; occupation of space with him was due entirely to substantial forces, not to the absolute hardness of primitive particles; and all strictly mechanical action of the push and press kind was replaced by attractions or repulsions acting at a distance. The kinetic theory can be regarded as a synthesis of these contraries. There is no action at a distance; but then there is no empty space: action and reaction are to be explained, not by impact, but by the physical continuity of the plenum. There are no hard atoms; yet the atom occupies space and is elastic in virtue of its rotatory motion.

Faraday, who has been called a disciple of Boscovich, made the *first* step on in the course of his wonderful electrical researches. He shewed that in the part of space traversed by magnetic force there exists a peculiar tension; as Maxwell puts it, "that wherever magnetic force exists there is matter"—that is to say, an electromagnetic medium or ether. Again Faraday's discovery of the magnetic rotation of the plane of polarised light, together with Maxwell's identification of the rate at which light and electro-magnetic disturbances are propagated. confirmed as this has been by the crucial experiments of Hertz, makes it reasonable to identify the lumin ferous and electro-magnetic media. The *second* great step towards this new ideal begins with the mathematical investigation of Helmholtz into the properties of vortex motion. Though apparently not suggested by Faraday's work, the two were soon brought into connexion; for Lord Kelvin found that the medium when under the action of magnetic force must be in a state of rotation, that is to say, in Maxwell's words "small portions of the medium, which we may call molecular vortices, are rotating, each on its own axis, the direction of this axis being that of the magnetic force."[1] *Finally*, Helmholtz's demonstration of the conservation of vortex-motion in a perfect fluid led Lord Kelvin to his famous vortex-atom theory, of which I have already spoken, and which in its main features is known to everybody. According to the kinetic ideal of matter, then, both atoms and ether are resolved into motions of one ultimate fluid, which is defined as having "no other properties than

[1] *Scientific Papers*, vol. ii, p. 321.

inertia, invariable density, and perfect mobility; and the method by which the motion of this fluid is to be traced is pure mathematical analysis."[1]

Let me quote two versions of what is expected of this ideal from two of its most able and hopeful supporters. Dr. Larmor, in a paper in the Royal Society's Proceedings of 1893, writes: "It has been in particular the aim of Lord Kelvin to deduce material phenomena from the play of inertia involved in the motion of a structureless primordial fluid; if this were achieved it would reduce the duality, rather the many-sidedness, of physical phenomena, to a simple unity of scheme; it would be the ultimate simplification." This brief statement is clear and modest by comparison with the following deliverance of Professor Hicks in his *Address* to Section A at the last meeting (1895) of the British Association: "While on the one hand," said Professor Hicks, "the end of scientific investigation is the discovery of laws, on the other, science will have reached its highest goal when it shall have reduced ultimate laws to one or two, the necessity of which lies outside the sphere of our cognition. These ultimate laws — in the domain of physical science at least — will be the dynamical laws of the relations of matter to number, space, and time. The ultimate data will be number, matter, space, and time themselves. When these relations shall be known, all physical phenomena will be a branch of pure mathematics. We shall have done away with the necessity of the conception of potential energy, even if it may still be convenient to retain it; and — if it should be found that all phenomena are mani-

[1] Maxwell, *o.c.*, vol. ii, p. 471.

festations of motion of one single continuous medium — the idea of force will be banished also, and the study of dynamics replaced by the study of the equation of continuity."

Every sentence in these remarks would repay criticism, if we could spare the time. As it is, I must content myself with an occasional reference in the more general criticism of this ultra-physical ideal to which we may now pass. But first, I will ask your indulgence if I quote part of yet another paragraph from this presidential address. "Before, however, this can be attained," Professor Hicks continues, "we must have the working drawings of the details of the mechanism we have to deal with. These details lie outside the scope of our bodily senses; we cannot see, or feel, or hear them, and this, not because they are unseeable, but because our senses are too coarse-grained to transmit impressions of them to our mind. The ordinary methods of investigation here fail us; we must proceed by a special method, and make a bridge of communication between the mechanism and our senses by means of hypotheses. By our imagination, experience, intuition we form theories, we deduce the consequences of these theories on phenomena which come within the range of our senses, and reject or modify and try again. It is a slow and laborious process. The wreckage of rejected theories is appalling; but a *knowledge of what actually goes on behind what we can see or feel* is surely if slowly being attained."[1]

Now I think the whole drift of these statements, and particularly this last sentence, makes it abundantly plain

[1] *Nature*, vol. lii, p. 472; italics mine.

that Dr. Hicks — and I am sure he is not alone — regards the hydro-kinetic theory of matter which he passes on to discuss, *not* as so much descriptive parable or 'conceptual shorthand,' but as veritable, conceivably perceptible, reality; in short, "what actually goes on behind what we can see or feel." Very good. Let us now try to understand what this means.

If this primordial fluid is real, it must have some positive attributes, and it cannot be an abstraction. But it is defined as inert, incompressible, inextensible, inviscid, and structureless, all negative terms. It is useless to reply that it is quite indifferent whether we use words that are positive, or words that are negative in form; that, in fact, this primitive fluid can be equally well defined as massive, of constant density, perfectly mobile, and absolutely homogeneous and continuous. Leaving the question of mass or inertia aside for a time, — we shall have to deal with it more at length, presently, — the remaining properties are, I take it, all summed up in the one phrase 'perfect fluid.' And as all the fluids we know are *im*perfect, it might seem that the negation belongs to the known, not to the unknown. But to say nothing of the obvious impossibility of this, we find that the characteristics of an imperfect fluid, one and all, refer to experimental facts. All liquids are compressible, viscid, and more or less discrete or structural. Let me cite a witness who has some claim to speak on such a point, I mean Clifford: — "A true explanation describes the previous unknown in terms of the known; thus light is described as a vibration, and such properties of light as are also properties of vibrations are thereby explained. Now a perfect liquid is not a known thing, but a pure fiction. The imperfect liquids which approximate to it, and from which the conception is derived, consist of a vast number of small particles perpetually interfering with one another's motion. . . . Thus a liquid is not an ultimate conception, but is explained — it is known to be, made up of molecules; and the explanation requires that it should not be frictionless. The liquid of Sir William Thomson's hypothesis is continuous, infinitely divisible, not made of molecules at all, and it is absolutely frictionless. This is as much a mere mathematical fiction as the attracting and repelling points of Boscovich."[1] Even Professor Lodge, though a sturdy upholder of the hydro-kinetic ideal, seems willing to allow the impropriety of the term 'fluid.' "Ether," he says, "is often called a fluid or a liquid, and again it has been called a solid, . . . but none of these names are very much good ; all these are molecular groupings, and therefore not like ether [the name Professor Lodge applies to this primitive medium] ; let us think simply and solely of a continuous frictionless medium possessing inertia, and the vagueness of the notion will be nothing more than is proper in the present state of our knowledge."[2]

Very good; again leaving aside for a moment the property of inertia, let us think simply and solely of this "continuous frictionless medium," neither ordinary fluid nor solid. Wherein does it differ from space? Space too is incompressible, inextensible, frictionless, and structureless, and it furnishes the very form and type

[1] *Lectures and Essays*, vol. i, p. 238 f.

[2] *The Ether and its Functions*, *Nature*, vol. xxvii, p. 305.

of a continuous medium. But whereas space is a perfect vacuum, it will be replied, our medium is a perfect plenum. But from empty space to masses in motion is a distinct step and from a uniformly filled space the step is just as distinct. So far as the realisation of any form or motion, thing or process, is her one aim, Nature ought to abhor such a plenum quite as cordially as she is said to abhor a vacuum. But the primordial medium has mass, we shall be reminded; in other words, it is inert, and inertia at least is a definite and fundamental physical fact. Let us now, then, inquire whether this remaining attribute of the universal medium renders it any more determinate, or whether, as so applied, 'inert' is anything better than another negation.

Inertia as a qualitative term and in its primary sense of inability or incapability is obviously negative. So Young defined inertia as the incapability of matter to alter its existing state except under the influence of some external cause. To allow that this universal plenum has inertia then does not remove its indeterminateness. Before it can be determined or differentiated in any way, some cause must intervene entirely from without, and such intervention will not admit of physical description. Such cause is of the nature of creation or miracle; it is neither a force in the sense of the attractions or repulsions by which Boscovich and Kant sought to explain matter, nor is it force in the modern sense of mass-acceleration. In other words, in the kinetic ideal of matter we shall find that the notion of mass is used with two distinct and inconsistent connotations. Abstract mechanics, as we have seen, sets out from definite masses or bodies having assignable positions, between every two of which there are dynamical transactions. *Two* masses, that is to say, measure each other by their mutual accelerations; in other words, mass is a strictly quantitative notion, and as such implies relation to a standard. Not only is mass in this wise always a relative quantity, but it is relative again in implicating the correlative notion of moving forces or stress between masses, which, as just said, is the only means of determining mass. If we attempt to apply the notion of mass to a universal homogeneous plenum, it lapses back into the merely qualitative notion of incapability of change evenly diffused through all immensity. And definite forces — necessarily present where there are definite masses to interact — seem here excluded. I trust I am not mistaken on this point. But it is difficult to imagine what definite forces there can be. Everything chemical or thermal or electrical is excluded, for the medium is throughout homogeneous and structureless. In like manner gravity, elasticity, and cohesion seem incompatible with absolute inviscidity and uniform density. Accordingly, to secure stability, it has to be postulated either that the medium is provided with a fixed boundary or that it extends to infinity. Mathematically these alternatives may come to the same thing, though the latter, *i.e.* infinite extension, seems the simpler and less arbitrary of the two, again shewing how little there is to choose between a vacuum and this plenum. The properties of such a plenum, indeed, as Maxwell chanced to remark[1] a year before Lord Kelvin's great

[1] *Scientific Papers*, vol. ii, p. 26 (on *Dynamic Theory of Gases*).

hypothesis was broached, "may be dogmatically asserted but cannot be mathematically explained." The reason for this seems simple: such a medium does not furnish even to abstract mechanics any πoῦ στῶ.

However, assuming that in some ultra-physical fashion it has been whisked up into that state of turbulent motion to which Lord Kelvin has given the name of "vortex-sponge," — this being the first step in cosmic confectionery, — let us see how this primitive mass is related to the phenomenal masses that then appear. The point I wish to urge is that neither the one nor the other conforms to the conception of mass with which abstract mechanics set out. The mass of every portion of the primitive fluid is an inalienable property of that portion. So far good, of course. Again, since the fluid is, and ever remains, of uniform density, the primitive or 'actual mass' of every portion is proportional to its volume. A vortex-ring is such a portion. But now its mass as measured by its mechanical effects is not simply proportional to its volume; in determining this 'effective mass,' the 'strength' of the vortex, *i.e.* its rotational motion, is also a distinct and independent factor. In short, this quasi-mass, or "non-matter in motion," depends upon a number of conditions, of which the real or primitive mass is only one. Such quasi-mass is therefore not an inalienable property in the sense in which primitive mass is such. For instance, though the volume of a vortex is constant, and therefore its primitive mass also, its configuration is liable to vary — in which fact of course lies the chief merit of the vortex-atom. But on these variations in its configuration depends the extent to which other portions of fluid are carried along with the vortex, as it moves onwards. Thus, while its primitive mass is invariable, its effective mass may vary with its motion and configuration.

We are brought, in short, to this paradoxical result: First, mechanical mass, the mass we know, is resolved into a mode of motion of some ultra-physical mass not directly capable of mechanical transactions, a mass that we therefore do not, and cannot, know as such. Given so much space, there is given also so much of this ultra-physical mass; but how much or how little nobody can say. Our scientific teachers have trespassed unawares beyond the limits of the phenomenal, and we find ourselves bowing down to a 'fetish' after all, none other indeed than that hoary idol of metaphysics, *τὸ ἄπειρον*, *materia prima*,[1] qualitatively indeterminate and quantitatively indistinguishable from space. Secondly, a mechanical, effective, or apparent mass, instead of being a constant and ultimate physical quantity, as at first defined, proves, so Professor Hicks tells us, "a much more complicated matter, and requires much fuller consideration than has been given to it." It may even, he thinks, "depend to some extent at least on temperature, however repugnant this may be to current ideas." Thus in this endeavour to carry through the application of abstract mechanics to all physical phenomena, the conception of mass proper has got pushed over the brink of the sensible and empirically verifiable, and seems in danger of being lost in those terrible

[1] Cf. Descartes, *Les Principes de la Philosophie*, bk. ii, art. 5.

'metaphysical quagmires' at which, as we have seen, the reputable physicist shudders. So now, instead of having this conception to the good in explaining or describing physical phenomena, the *semblance* of mass has itself to be accounted for; and this, as we have just been told, is a very complicated business "requiring much fuller consideration than has been given to it." The *impasse* which thus threatens to end the kinetic ideal of matter was clearly seen by Maxwell and is admitted by Lord Kelvin. In the article '*Atom*' in the *Encyclopædia Britannica* Maxwell thus criticises it: "Though the primitive fluid is the only true matter, according to the kinetic ideal that is to say, yet that which we call matter is not the primitive fluid itself, but a mode of motion of that primitive fluid. . . . In Thomson's theory therefore the mass of bodies requires explanation. We have to explain the inertia of what is only a mode of motion, and inertia is a property of matter, not of modes of motion." Lord Kelvin himself, in concluding his lecture on '*Elasticity as a Mode of Motion*,' acknowledges that "this kinetic theory of matter is a dream and can be nothing else, until it can explain," not only the "inertia of masses (that is, crowds) of vortices," but also gravitation, chemical affinity, and much besides. His only ground of confidence appears to be the "belief that no other theory of matter is possible."[1] But this was in 1881; and one cannot help wondering whether Lord Kelvin's confidence in his theory has increased or diminished in the meantime. Some among the younger generation of physicists pre-

[1] *Popular Lectures*, vol. i, p. 145.

fer, as I mentioned in the last lecture, to abandon the attempt to reduce all physical phenomena to a connected mechanism based solely on the Newtonian laws. Many of them look to find a better way by taking, not mass, but energy, for the fundamental notion. Before we pass on to this, however, it will be well to try to gather up the main results of our inquiry into the mechanical theory so far.

We have distinguished three branches of science which, though distinct, are closely connected and often confused: (1) Pure, or Analytical Mechanics; (2) Mechanics applied to Molar Physics, which might be called Molar Mechanics; and (3) Mechanics applied to Molecular Physics, or Molecular Mechanics. The first is in the strictest sense an exact science based entirely on certain fundamental assumptions and definitions. We have here rigorous calculation, but no measurement: ideas, but not facts. The other two rest in part on observation and experiment, which yield approximate measurements, probable values, *i.e.* averages and means corrected by the help of that — for the student of knowledge — most wonderful instrument, 'the logic of chance.' In the exact sciences, within the limits of our powers and subject only to the laws of thought — we are complete masters of the situation. Our intellectual constructions are archetypal and not ectypal. We can here give a meaning to absolute time, absolute space, absolute motion; we can here talk reasonably of the perfectly continuous, perfectly discrete, and perfectly constant. But applied to the particulars of experience such conceptions have no warrant. The Pythagorean proposition, for example, is exact and certain, apart from all physical circumstances as a proposition in plane geometry. But, as Riemann's famous dissertation suggests, it is quite conceivable that this proposition should be falsified one way in astronomical measurements, if the distances measured were sufficiently vast; and be falsified another way — in mineralogical measurements, say — if these distances were sufficiently minute. Of course we might prefer to consider our lines as not really straight. But this might quite well only be changing one contradiction for another, or prove far less simple than it would be to describe the facts in terms of some non-Euclidean space. But worse than this and far less open to dispute: the most elementary conditions of absolute exactness everywhere fail us. We have no fixed points, no fixed directions, no accurate timekeeper, not one *demonstrably* constant property of a physical description. Even number when applied to physical phenomena is no exception, in so far as neither identity nor simplicity nor discreteness admit of more than a relative application.

Now, as a consequence of all this, if you like—as the price of its formal exactness, abstract mechanics has to renounce those higher categories, Substantiality and Causality, which bring us into touch with concrete things. The process of eliminating these categories has been slow; for the terms 'mass' and 'force' seem almost inseparably associated with substance and power, from which notions in fact they were primarily derived. But regarding the result as at length complete, and accepting the purely mathematical definitions of mass and force now in vogue, the bearing of this result on molar and molecular mechanics is important. The simplest and most comprehensive description of the movements, actual or supposed, that occur in nature becomes the sole aim of these sciences, not the unveiling of the mystery of matter or the knowledge of the causes of things. The logical development of this procedure we have attempted to follow in some detail, and the outcome, as we have just seen, is that we find nothing definite except movement left. Heat is a mode of motion, elasticity is a mode of motion, light and magnetism are modes of motion. Nay, mass itself is, in the end, supposed to be but a mode of motion of a something that is neither solid nor liquid nor gas, that is neither itself a body nor an aggregate of bodies, that is not phenomenal and must not be noumenal, a veritable *ἄπειρον* on which we can impose our own terms. I am sure this process will remind many of you of one of *Alice's Adventures in Wonderland.* I trust I may be pardoned for the allusion. The Cheshire Cat, you remember, on a certain occasion, "vanished quite slowly, beginning with the end of the tail and ending with the grin, which remained some time after the rest of it had gone. 'Well! I've often seen a cat without a grin,' thought Alice, 'but a grin without a cat! It's the most curious thing I ever saw in all my life.'"

In this advance towards what looks like physical nihilism, molar and molecular mechanics constitute each a distinct step. The salient feature we have noted in molar mechanics is that 'species of abstraction' that Thomson and Tait describe as 'limitation of the data.' Of such abstractions we have an instance in the treatment of the constraints and connexions that limit the free motion of a particle or of the separate portions of a machine, as mere geometrical or kinematic conditions. In actual fact constraint involves friction, strings stretch, levers bend, and so on. But all these imply intermolecular forces, the investigation of which is passed on to experimental physics. Again a change in the momentum of a body may be due to any one or more of a variety of causes—gravitation, heat, chemical action, and so on. Molar mechanics considers none of these: it is concerned only with the rate of the change itself, giving, as we must remember, the name of 'moving force' to this *effect*. The various causes, as we are allowed provisionally to call them, are, as before, passed on to corresponding departments of experimental physics. Finally the bodies moving have manifold properties. Of these all save mass and mobility are ignored, and the rest again passed on to experimental physics.

But now assume for a moment that molecular mechanics has fully accomplished the task assigned to it, I mean this mechanical interpretation of the facts of experimental physics. None of those conditions of constraint, none of those natural forces or physical properties, which molar mechanics passed on, will then be left over; all of them will have been described in terms of mass and motion. It is thus obvious that that 'species of abstraction' or limitation so characteristic of the methods of molar mechanics does not pertain to molecular mechanics. On the contrary, that science, if verily complete, would — we have been told — embrace in one scheme all the vast variety of physical phenomena reduced to the simplest possible form. True, its fundamental ideas would be the same as those of pure mechanics, but then we should be assured that there were no others, whereas in molar mechanics this still remained an open question. In fact this last science would itself be absorbed; inasmuch as a body of sensible dimensions would be but an aggregate of molecules, and all those of its properties, left aside as non-mechanical in the aggregate, would be referred to mechanical processes in the parts. It is allowed, of course, that molecular mechanics is *not* complete; and we have seen that its procedure, when seeking to express the facts of chemistry, light, electricity, etc., in purely mechanical terms is in the main hypothetical and indirect. Molecules, Atoms, Ethers, *Prima Materia* — one and all are hypothetical. "Nevertheless," say the naturalists, "they are thoroughly sound hypotheses and their scientific value is enhanced daily both by known facts that they are continually assimilating, and new facts that they are continually revealing. We realise that there is still much to do, but at the same time we are confident that 'no other theory of matter is possible.' Our scheme is therefore regarded as established in principle despite important gaps in detail."

Now it is this advance — from mechanical theory, as a branch of pure mathematics, through molar mechanics, as an abstract application of that theory, on to molecular mechanics, in which all physical phenomena are subsumed under it — that vitally concerns us. A science which at the outset is simply formal and quantitative seems in the end to yield the ideal of concrete physical existence, what Kant might have called the *omnitudo realitatis* of the physical world; and this becomes, for those to whom the physical world is primary and fundamental, the supreme and only *omnitudo realitatis* that science can ever know. Here, then, we have that advancing tide of matter which, as Huxley says, "weighs like a nightmare on the best minds of these days." But surely if our account of this transformation of pure mathematics into concrete physics is correct, the baleful spectre should be dispelled, and that without any recourse to such an agnosticism as Huxley's. The mechanical theory, in a word, as I have already hinted, refutes itself by proving too much. Or, to put it otherwise, and more fairly: the mechanical theory, as a professed explanation of the world, receives its death-blow from the progress of mechanical physics itself.

As long as the ideal of matter consisted of the "solid, massy, hard, impenetrable, movable particles of various sizes and figures" (such as Newton supposes in his *Opticks*), maintained in various states of vibration, rotation, and translation by their mutual encounters; so long this ideal of matter answers to Newton's conception of a *vera causa*. But the simple atom or centre of force of Boscovich, and the primitive fluid of Lord Kelvin, are not *veræ causæ*: we must not call them fetishes, but they are assuredly fictions. To Newton's particles we *might*, perhaps, apply Dr. Hicks's words: "They lie outside the scope of our bodily senses; . . . not because they are imperceptible, but because our senses are too coarse-grained to transmit impressions of them to our minds." To bodies wholly devoid of extension, or to a plenum wholly devoid of differences, such language cannot be applied. The process of analysis up to the stage of the chemical or physical molecule, though hypothetical and indirect, may yet be regarded as *real* analysis; and had the hypothesis of extended molecules proved adequate, the mechanical theory might, so far as science goes, have held its ground. Extended, solid, indestructible atoms have always been the stronghold of materialistic views of the universe. But, unhappily for such views, the hard, extended atom was not equal to the demands which increasing knowledge laid upon it. Then, as we have seen, encouraged by Newton's essentially descriptive conception of distance-action, the old atom shrank up gradually, surrendering all its extension, rigidity, and elasticity, till it became identical with the entirely formal conception of analytical mechanics, that, viz., of a mass-point as a centre of force. But this later analysis, though still hypothetical, had no longer any conceivable physical counterpart. The supposition that it had was due solely to that failure to realise the purely descriptive character of mechanics which its increasing mathematical formulation and its liberation from the categories of substance and cause have now made clear. It fell to Pere Boscovich decently to inter the genuinely mechanical theory as an explanation of physical phenomena. There was no rest for the old atom till it took this ghostly form of a mass-point, and thenceforward it was a dynamical fiction, pure and simple.

Lord Kelvin's brilliant hypothesis of vortex-atoms, if

regarded as an endeavour to resuscitate indestructible and extended atoms as realities, and to provide a medium for their interaction, must be pronounced a failure too. Boscovich resolved the palpable atom into an idea ; Lord Kelvin seems to attempt the converse and far harder feat of calling back this atom from a "vasty deep" so dangerously like pure being as to be, phenomenally, pure nothing. The endeavour to attribute mass to this continuum is as if one should let one's plummet drop in the hope of sounding a fathomless sea ; we lose a simple conception, and have a complex one left on our hands instead. But now comes Dr. Hicks to persuade us that we gain more than we lose : "If it should be found that all phenomena are manifestations of motion of one single continuous medium, the idea of force will be banished [the relative idea, that is, of which mass is the correlative] . . . and the study of dynamics will be replaced by the equation of continuity ;" for "where all the matter is of the same density the motions are kinematically deducible from the configuration at the instant, and are independent of the density."

These remarks are most opportune. If we consider them for a moment, they ought to satisfy us that we are *not* penetrating beyond what we see and feel to anything that actually goes on behind the too coarse-grained veil of sense. They serve to shew, on the contrary, that the kinetic ideal also is but a fiction of the mathematician, a descriptive symbol, and not conceivably a presentable fact. Now there is a certain philosophical doctrine, both psychologically and epistemologically of fundamental importance, that ought to be well known in Aberdeen[1] — I mean the doctrine of the relativity of knowledge. The range of this doctrine may be very much a question, but at least no one will deny that it applies here. See then to what it leads. Everything perceptually real, everything phenomenal, whatever can be an object of possible experience, implies difference and change. But we have left all sensible qualities except density behind us; and this, though retained, is to admit of neither difference nor change. "*Idem semper sentire et non sentire ad idem recidunt*," says the doctrine of relativity. For any conceivable experience then this density is as nothing. Moreover, according to the kinetic theory, the motions are independent of it. Why then is it retained? Apparently to stand between us and nonentity. It secures for us that "idea of *stuff* or *substance* which," Professor Tait tells us, "the mind seems to require" — well for comfort![2] It is then *das reine Sein* of our present universe of discourse. Or it is the 'Achilles heel' of reality, left when all the rest of the physical world has been dipped in the Styx. "But why," asked an intelligent child, "did not Thetis dip Achilles twice?" Now Dr. Hicks appears to have had that much foresight in agreeing to let go dynamics and to abide by the equation of continuity. For dynamics and mass must surely vanish together, and we have properly only kinematics left. Nevertheless there remains one stipulation that kinematics does not warrant — there must be no discontinuity in the

[1] Being so strenuously maintained by Dr. Bain.

[2] *Unseen Universe*, p. 105.

motions on two sides of a geometrical boundary. The vortices, in other words, must not spin and leave the medium unaffected; and so the medium, being involved in the movement of one vortex, must in turn affect the movements of another. And thus with this proviso the whole becomes, as we may say, one vast quasi-dynamical or rather quasi-kinematical system. For it is allowed,[1] I believe, that the existence of surfaces of finite slip is not precluded by the bare conception of a uniform frictionless medium. Imagine such an ideal fluid if you can, and the question whether a vortex in it will or will not affect the fluid outside the vortex is altogether indeterminate. It may do either or neither, sometimes the one and sometimes the other. Why then is this condition of motional continuity imposed from without? Simply to make the thing work *mathematically*, that is to say, to insure connexion and continuity between one kinematical configuration and another. Without it we might have vortex-atoms as before, but not "actions excited by these vortices on one another through the inertia of the fluid which is their basis."[2] Such mutual regard is not then a direct consequence of the common plenum. In fine, then, this additional property of motional continuity is asserted, though it cannot be deduced, in order to make possible a kinematical scheme that *replaces*, as Dr. Hicks says, the dynamical laws that can then be left behind.[3]

[1] See letter on *Vortex-atoms*, by Professor G. H. Darwin, *Nature*, vol. xxii, p. 95. [2] Dr. Larmor, Proc. R. S., 1893, p. 439.

[3] "It will be seen that the work is almost entirely kinematical; we start with the fact that the vortex-ring always consists of the same parti-

It may be that this exposition by the President of the Physics Section of the British Association sounded rash, or at least premature, to the distinguished physicists who heard it. But it must certainly be impressive to any humble outsider with a philosophical bent. It exhibits strikingly the complete logical outcome of the problem of mathematical physics, as formulated by the Kirchhoff school; and all the more strikingly because this consequence is here worked out, as it were unconsciously, by one who, unlike Kirchhoff, seems to suppose that he is all the while getting nearer to "what actually goes on" in the real world. The tendency to extend kinematics at the expense of dynamics seems inherent in this new conception of physics. But the sounder the conception, the more this tendency may be expected to assert itself spite of contrary prepossessions, and the more effectively will such prepossessions be dispossessed.

Now it is entirely upon these uncritical prejudices, as we may fairly call them, that the mechanical theory of the world rests. The more they are discredited the more it is discredited through them, and this, I believe, the history of science will amply show. The transference of motion by impact, for example, as when two billiard balls collide, seems the type of plainness, and so

cles of fluid (the proof of which, however, requires dynamical considerations), and we find that the rest of the work is kinematical. This is further evidence that the vortex theory of matter is of a much more fundamental character than the ordinary solid particle theory, since the material action of two vortex-rings can be found by kinematical principles, whilst the 'clash of atoms' in the ordinary theory introduces us to forces which themselves demand a theory to explain them." Professor J. J. Thomson, *A Treatise on the Nature of Vortex-Rings*, 1883, p. 2.

long as this and other equally familiar experiences were accessible to the imagination, it seemed still to retain its grasp of the real spite of 'the cloud of analytical symbols.' The triumph of the undulatory, over the corpuscular, theory of light, was a blow to such realism; for an imponderable ether was not easy to conjure up by imagination. Still, after all, waves are familiar and it was only the 'undulating agency'[1] that was obscure. But a severer blow overtakes us in what we might call the demolition of the chemical atom as an assured stronghold of the realistic imagination. And when both chemical elements and luminiferous ether are resolved into motions of a medium, 'the dynamics of which is not the dynamics of ordinary matter,'[1] realism seems fairly routed. But stranger still, imagination has become itself a traitor to mechanical realism — I refer, of course, to such ingenious mechanical analogies as those, for example, by which Maxwell succeeded in elucidating electromagnetism. Analogy is an important aid to description, though powerless to prove existence. Nevertheless, as I had occasion to remark in the last lecture, even the ablest men are apt to see more in analogy than this; and it speaks volumes for Maxwell's strength of intellect that, acute as he was in the discernment of helpful analogies, he seems never to have been led away by them. But it is a case in which there is safety in numbers. A thinker familiar with many analogies is less likely to be betrayed by them than a thinker whose mind is enchanted by one. Now Boltzmann, in an instructive paper

[1] Lord Salisbury, *Presidential Address*, British Association, 1894.

[2] Larmor, *Nature*, vol. liii, p. 4.

on the *Methods of Theoretical Physics* from which I have already quoted once or twice, gives many instances of surprising and far-reaching analogies that have been discovered within the last half-century between physical phenomena apparently quite unlike; as if nature had "built up the most diversified things after exactly the same pattern." "As the analyst dryly observes, the same differential equations hold for the most diversified phenomena." And no great wonder if the analyst previously made up his mind to see the most diversified phenomena merely as cases of motion, to be described in the simplest and most comprehensive manner. The logical goal of such a project, I conclude then, is — if I may so say — to minimise the inevitable '*matter*' of phenomena and to bring all the diversity possible under the 'form' of motion. This goal is already set before us in the kinetic ideal of matter, where dynamics is all but sublimated into kinematics. So much so indeed, I may remark by the way, that even the motion is absolute, and not merely relative motion; for every motion is strictly a motion of the medium, and this is infinite and all there is. Now, as soon as we are asked to entertain the notion of absolute motion, we may satisfy ourselves that we have left everything phenomenal behind us and are once again entirely in the region of the abstract conceptions of exact mathematics. And the medium itself, though infinite and all there is — nay, because of this, for it does not allow even the distinction of body and space — is indistinguishable from nothing. The whole ideal, it seems to me, *if it be meant to set before us what verily is and happens*, was refuted long ago by Leibnitz in the following sentences of the *Monadology* (§ 8): "If simple substances did not differ at all in their qualities, there would be no way of perceiving any change in things, since what is in the compound can only come from the simple ingredients, and if the monads were without qualities they could not be distinguished the one from the other, since also they do not differ in quantity. Consequently, a plenum being supposed, each place in any movement could receive only the equivalent of what it had before, and one state of things would not be distinguishable from another."

We smile at the critical simplicity while admiring the boldness of the Pythagoreans, according to whom, as Aristotle tells us, "Number is the essence of all things; and the organization of the universe, in its various determinations, is a harmonious system of numbers and their relations." Enough perhaps is known of the Pythagoreans and their tenets to shew that they had no pure science of number, but that such arithmetical knowledge as they had was encumbered by concrete and fanciful associations with numbered things. May we not apply the moral to the mechanical theories of the universe, and say that the more clearly the purely mathematical character of mechanics is realized, the more absurdly inadequate that theory becomes? A science that can only offer us as its ultimate scheme of the universe the inconceivable ideal of continuous motion in an unvarying plenum, is surely as incompetent as arithmetic or geometry to furnish a concrete presentment of a real and living world. Its essentially formal character has become increasingly evident with every improvement in its methods. Galileo and Newton made many experiments, and their works abound in diagrams; but I am not aware that either Lagrange or Laplace ever tried an experiment, while Lagrange is said to have boasted that his *Mécanique analytique* did not contain a single figure. This science, then, which has gradually rid itself of the categories of substance and cause, which works entirely with abstract quantities, expressing its conditions in equations and its results in equations, does not, and cannot, yield any direct knowledge concerning real things. When employed to describe them, its application is restricted absolutely to the one quantitative aspect with which it deals, — the motions of mass-systems. It has no scientific status except where such motions are either (1) given, or (2) inferred, or (3) assumed. In the first case its results, though necessary and exact in themselves, become at once hypothetical and approximate in their application; the ideal simplicity and abstract isolation of theory being never found in reality. In the second case the results are more hypothetical and approximate still; for neither the particles nor the motions themselves can be directly measured. This is the region of statistical probabilities. In the third, the masses and motions are *entirely* hypothetical; it is no longer, strictly speaking, a case of applying pure mechanics to describe real motions. This is the region of mechanical analogies, of prime atoms and ethers, vortices and primordial fluids; the region in which, as Dr. Hicks has told us, "the wreckage of rejected theories is appalling."

The mechanical theory of the universe, then, begins

with abstractions, and in the end has only abstractions left; it begins with phenomenal movement and ends by resolving all phenomena into motion. It begins with real bodies in empty space, and ends with ideal motions in an imperceptible plenum. It begins with the dynamics of ordinary masses, and ends with a medium that needs no dynamics or has dynamics of its own. But between beginning and end, there are stages innumerable; in other words, the end is an unattainable ideal. First, we have sensible mechanisms; to these theoretical formulæ only apply approximately, their abstract simplifications being inadequate to cope with the 'practically infinite' complexity of the reality. A closer approximation is secured, but at the cost of new residual discrepancies, by resolving the parts of sensible mechanisms into smaller mechanisms, and the parts of these into others yet smaller in turn. Again, further approximations are made by attributing other elements of the real complexity to imaginary mechanisms of many orders. But the complexity being, as said, 'practically infinite,' this procedure has no prospect of ending. Dr. Hicks, for example, even when he has got as far as the chemical atom,—and that, we must remember, is a very long way,—cheerfully tells us, "The atom is much larger than a cell, and contains, practically, an infinite number of them;" a cell, I must tell you, being an imaginary box that Dr. Hicks has devised, in which a vortex of the primary medium is magically penned up to wriggle. Yet, spite of these complex mechanical fictions, *no* advance is yet reported towards a kinetic theory of gravitation, and very little has been done with the terrible complications of chemical affinity. The story of the progress so far is, then, briefly this : Divergence between theory and fact one part of the way, the wreckage of abandoned fictions for the rest, with an unattainable goal of phenomenal nihilism and ultra-physical mechanism beyond. Nevertheless, there are many who hold that the world must be such a mechanism, because they imagine themselves unable to conceive it otherwise. Such, as I understand it, is Lord Kelvin's position, for example. Others see in the situation a parallel to that of the Ptolemaic astronomy, which could not cope with increasing knowledge even with the help of new eccentrics and epicycles, freely assumed as the occasion arose. A new and simpler science of energetics is with some of these reactionaries the counterpart of the Copernican astronomy, and is to release physics from the complications in which mechanics has involved it. These are points that must occupy us in the next lecture.

LECTURE VI

THE THEORY OF ENERGY

The proposal to replace Mechanical Physics by Energetics. Whatever it may be worth, this proposal at least puts Mechanical Physics anew upon its trial.

I. What is energy? *Professor Tait's definition of Matter as the 'vehicle or receptacle of Energy' examined. Relation of Energy to Matter. Helmholtz's exposition of this relation. Relation of Energy to Mass. Is not Mass as much an analytical abstraction as Force?*

All change either a transference or a transformation of Energy, and Kinetic Energy only one *form of actual energy — this is the new doctrine. Difficulties of the old theory which is bent on resolving all actual energy into kinetic energy. Professor Duhem's protest, and some reflections that it suggests.*

Returning to the new theory we note (i.) that quantitative equivalence not qualitative identity is all that is asserted of the several forms of energy; and (ii.) that some of these forms may still remain undiscovered. Some final reflections on the mechanical bias.

II. What is the Conservation of Energy? *What it is not; it does not warrant statements about the past or future of the universe. It does not mean that Energy is verily and absolutely the substance of the universe. Its relativity. Its character as a postulate. Implications of this, and new questions opened up.*

IN the preface to the *Principia*, it will be remembered, Newton gave expression to his hope that if the mechanical principles he had laid down should prove inadequate to the explanation of "the other phenomena of nature, they might at least afford some light to a more perfect method of [natural] philosophy."[1] The inquiry which has occupied us for the last two lectures seems to shew that the first alternative is well-nigh, if not quite, hopeless. In place of simplifications of actual phenomena it offers us fictitious mechanisms; or mechanical analogies, in which quasi-rigidity, quasi-elasticity, quasi-mass, and quasi-matter meet us at every turn. One recent writer, the brilliant German physicist, Hertz, did not shrink from assuming that the underlying mechanism, by which he proposed to explain the effects we perceive, consists of hidden masses and motions that exceed by an infinite number the masses and motions to be described.[2] And even with all this more than poetic license it has not been found possible to resolve electrical and chemical phenomena into motions, to say nothing of the phenomena of organic life. Yet all these phenomena, it is said, are clearly amenable to the principle of the conservation of energy. Spite of the physicist's complete ignorance as to what the mechanism of electricity, for instance, may be, if indeed it has any mechanism at all; 'electric current' can be produced, measured, and retailed to consumers like other commodities; and is so far under control that it can be transformed into its equivalents of heat, light, or motive power. Nay, but for a knowledge of these transformations and their mechanical equivalents, the mechanical treatment of physics could not have advanced as far as it has. Here then is a principle universal in its range, independent of atomic hypotheses and fictitious forces, confirmed by innumerable experiments and contradicted

[1] Cf. Lecture III above, p. 84. [2] *Principien der Mechanik*, § 664.

by none, a principle that verily brings all physical phenomena, mechanical as well as the rest, under a single *real* scheme, surely this, it is said, is the true integral law of the world. And so just forty years ago Rankine sketched "the outlines of the science of energetics." The project has never been lost sight of, and within the last few years it has been pursued with ardour in many quarters, especially in Germany and in France. The views of the extreme upholders of this new science are still *sub judice*, so much so that it would ill become me as a complete layman in such questions to venture any opinion. But the doctrine of energy is fully admitted even by those physicists who are not prepared to yield it precedence over the old Newtonian mechanics. At the same time the more progressive doctrines are at least effective as criticisms of the older view. They are a new outgrowth, which, if it does not displace, must at least profoundly modify the older form. For these reasons it has seemed to me best to reserve the discussion of this subject till now, and to do so was easy, as the mechanical ideal contrives to dispense with all forms of actual energy save the old *vis viva*. And let me remark, by the way, that energy, as I understand, is to be regarded as a physical fact and not as a mathematical conception; in discussing it and the criticisms of the mechanical theory that it suggests, we are not then concerned with abstract mechanics as a branch of mathematics, but only with mechanics applied to physical phenomena.

This becomes evident when we ask: What is energy? It is in the answer to this question that we come upon the new wine that is to try the old bottles of the mechanical theory; for energy is so defined as to threaten the independent existence of that matter which was first of all regarded as its necessary substratum. Thus Professor Tait informs us that "in the physical universe there are but two classes of things, Matter and Energy." Further, that as "energy is never found except in association with matter . . . we might define matter as the *Vehicle* or *Receptacle of* Energy."[1] *Vehicle*, I presume, we are to take as the appropriate simile where the energy is actual and changes are in process; *receptacle*, when the energy is only 'stored,' and changes are only potential. But either way these figurative expressions distinctly imply that we know by experience each of these two things, just as we know and distinguish the cycle and the rider, the basket and its contents. The appropriateness of such language turns entirely on the question whether or no we have such knowledge. It will not do to say: We *must* have it, since we know that both matter and energy are conserved. We shall come to that presently; but it is plain our knowledge cannot begin there. To know such laws *about* the things, we must first have some sensible acquaintance with the things themselves. We get a little nearer to what we want when Professor Tait goes on to say: "Matter is simply passive (*inert* is the scientific word); energy is perpetually undergoing transformation." But surely to be perpetually *undergoing* transformation is no better than the dreariest picture of unmitigated passivity. However, Professor Tait continues: "the one

[1] *Properties of Matter*, pp. 2, 4.

[matter] is, as it were, the body of the physical universe; the other [energy] is its life and activity."[1]

Our question, then, can now be more precisely put; it is not, What do we or what does Professor Tait know *about* this simply passive thing, this inactive unchangeable body, as it were; 'scientific words' like inert, conservation, and the like, being used. The question is: What sensible acquaintance have we with the thing itself? Now it is remarkable that, although the book I have been quoting is entitled *Properties of Matter*—Professor Tait proceeds to say: "From the *strictly* scientific point of view the greater part of the present work would be said to deal with energy rather than matter;" and he only justifies the title he has used on "the two grounds of custom and convenience." We are not, however, concerned either with custom or with the convenience of exposition: on the contrary, it *is* the "strictly scientific" answer we want to the question: How far matter can be known apart from energy? The answer is: It cannot be known at all. I do not give this as the answer of philosophers, it is the answer of the physicists themselves. Every physical quality we distinguish, every physical change we observe, every physical measurement or comparison we can make, relates to energy, to the "life and activity" of the physical universe; not one refers to the supposed vehicle or receptacle, "the body, as it were," of that activity. In that famous memoir on the subject, which fifty years ago was rejected as nonsense, though it has now become one of the corner-stones of the new edifice, Helm-

[1] *Properties of Matter*, p. 5.

holtz concedes this point, without however realizing its consequences. The point is one which you may perhaps think I have laboured at sufficiently already when endeavouring to make clear the extremely abstract nature of the conception of mass. But we are approaching it now from what we might call the opposite side, and I am anxious that at every stage we should keep our authorities well in sight. Let me then quote a few sentences from the philosophical introduction, as he calls it, with which Helmholtz prefaced his essay on *The Conservation of Energy*. And please note that what primarily concerns us is the answer that his words afford to our question as to the possibility of knowing matter as distinct from energy. His philosophy of the relation of the two conceptions we can examine later. "Science," he tells us, then, "deals with external objects from two abstract points of view: first, as barely existent, apart from their effects on other objects or on our organs of sense; as such we call them matter, which for us is a thing in itself without motion, without action. Qualitative differences are not to be ascribed to it, for so soon as we speak of different kinds of matter we imply differences of operations, *i.e.* of energies. Natural objects however are not without action, for we become acquainted with them at all solely through their actions, by which they eventually affect our senses; while from these actions we infer a something acting. In applying the conception of matter, therefore, to actual things, we must restore through a second abstraction what we were previously for leaving aside, the power to produce effects; in other words, we must assign it energy. It is manifest that, when applied to nature, the conceptions of matter and energy are not to be separated. Pure matter would for the rest of nature be a thing of indifference, since it would never determine any change either in this or in our senses. Pure energy would be something that ought to exist and yet again ought not to exist, for the existent we call matter. . . . Both conceptions are abstractions from the actual formed in the same way; we can in truth perceive matter only through its energies, never in itself."[1] Now here we have the most unequivocal admission that of matter as the simply passive vehicle or receptacle of energy we perceive nothing; that all we perceive of external objects is due wholly and solely to energy and to energy alone. True, Helmholtz proposes to treat both matter and energy as abstractions that are on the same footing; but in so doing, though—like Professor Tait afterwards—he conforms to custom and convenience, he flies straight in the face of the strictly scientific view. No doubt we *call* matter the existent, attributing energy as a power or property to it, and attributes cannot be separated from their substances. But great as are the forces of custom and the claims of established conventions, all that the facts lead us to infer is a "something *acting*," not a something passive, which would be a thing of indifference for everything beside. To the one conception corresponds, in short, all our perceptual experience; to the other the unutterably metaphysical notion of bare existence

[1] *Ueber die Erhaltung der Kraft*, p. 4, Ostwald's Klassiker der exakten Wissenschaft. 'Kraft' translated 'energy' throughout.

per se. How then can they be both on the same footing, especially for a scientific view that discards the notion of substance as non-phenomenal and defines matter as "that which can be perceived by the senses?"[1] Energy and its transformations are given, and nothing else is given; those who wish may attach the idea of substantiality or actuality to this, but they may not multiply entities needlessly. It would seem, then, that there are *not*, after all, two classes of things in the physical universe, but one only. Such at least appears to be the logical outcome of the theory of energy.

But what of mass, it will be asked; surely mass is a property of matter, is, in fact, that very passivity which distinguishes matter from energy. To answer this we have only to ask another question: Is mass perceptible by an external sense or is it not? Now, if we turn to our text-book, Professor Tait tells us first of all that "the mass of a body is estimated by its inertia;" next that inertia "may be described as passivity or dogged perseverance" in the motor *status quo*, "familiar instances of which present themselves in all directions," as when the "sudden stopping of a train appears to urge the passengers forwards."[2] In other words, we become acquainted with inertia when we experience a change of momentum, and in no other way. But such an experience, whether we regard it as a change or as a perseverance against change, implies time and implies 'the action of natural objects.' How are we going to advance from this to mass as pure passivity, which implies neither?

[1] Thomson and Tait, *Natural Philosophy*, p. 207.

[2] *Properties of Matter*, pp. 91 f.

To that we can find no answering experience. But we have seen how theoretical mechanics by analysing the dynamical transactions in which momentum is changed has reached the two abstract conceptions of mass and force. That the latter of these terms is nothing but an analytical abstraction Professor Tait has taught us with commendable emphasis and persistence. Is it not then odd that he is so anxious to persuade us that the former is a reality? Surely here at least the two abstractions *are* on the same footing. Then must we not decline to accept masses as *things*, in which energy careers like Ariel "on the curl'd clouds"; or between which it is imprisoned, like Ariel "'twixt a cloven pine"?

All change is either a transference or a tranformation of energy — this is the new doctrine. The familiar experiences to which we owe the conception of inertia are transferences of one particular form of energy, viz., motional or kinetic energy. This energy of motion may be mathematically regarded as $\frac{momentum \times velocity}{2}$ or, as Clifford once put it, half the rate at which momentum is carried along.[1] It is now, of course, a familiar fact that other forms of energy have their *equivalents* in kinetic energy and *vice versa;* it is this fact, indeed, that renders the doctrine of energy physically so important. But it is not a fact that other forms of energy are not only quantitatively commensurable, but qualitatively identical, with energy of motion. This qualitative identity is at best but an assumption; and in the vain

[1] *Nature*, vol. xxii, p. 123.

endeavour to justify it we have seen the mechanical theory led, "to pass through the very den of the metaphysician strewed with the remains of former explorers, and abhorred by every man of science."[1]

It is this instinct of self-preservation that prompts so many physicists just now to abandon as 'foolhardy' the adventure of mechanical physics, and to set about the construction of what we might call energetical physics instead. Let me quote one of them, Professor Duhem of Lille. Referring to the mechanical method, and after illustrating its futility in chemical physics, he says:—"We have seen this method at work; we have ascertained to how small an extent experience accords with the results of its deductions. In the face of such rebuffs is it not prudent to renounce the doctrines followed thus far? Why seek by mechanical constructions to set aside bodies and their modifications, instead of taking them as our senses give them, or rather as our abstracting faculty, working on the data of sense, leads us to conceive them? . . . Why seek to figure changes of state as displacements, juxtapositions of molecules, variations of path, instead of characterising such changes of state by the disturbance introduced into the sensible and measurable properties of the body, such, *e.g.* as increase or decrease of density, absorption or evolution of heat, etc.? Why wish that the axioms on which every theory must rest should be propositions furnished by statics or dynamics, instead of accepting for principles laws founded on experience and formulated by induction, whatever be the form of such

[1] Maxwell, *Collected Papers*, ii, p. 216.

laws and whatever be the nature of the concepts to which they appeal?"[1]

Such language as a protest against the *intellectus mathematicæ permissus* sounds like the counterpart to Bacon's against the *intellectus sibi permissus*, and leads one to wonder whether, after all, one and the same infirmity will not account for both — I mean that hankering after certainty and definiteness by which we are hurried into hasty generalisations. It was this that Bacon exposed as the anticipation of nature, while ironically praising it as so much easier and more satisfying a method than the patient interpretation of nature. It was to this too that Descartes referred when he declared the will and not the intellect to be the source of errors. A mechanism may be very complex, but once get at the working drawings, and then, as Professor Hicks suggests, there are no surprises, no irregularities, no uncertainties; only master the mathematics, and you are intellectually master of the whole. That is one reason why so many "wish that the axioms on which every theory must rest should be furnished by statics or dynamics." And there is another reason still, and one to which even Descartes, spite of all his rules, completely succumbed — I mean the influence of the imagination. We figure changes of state as being displacements or motions because we can imagine nothing else with equal clearness and distinctness. We cannot be surprised then that the certainty of mathematics, and the freedom from contradiction and obscurity of mechanical imagery, should have led so many able minds to an anticipation of nature that is

[1] *Méchanique chimique*, 1893, p. 88.

unwarranted by facts, and even induced them to affirm as Descartes, yes, and Kant too, have done, that a true science of nature extends just as far as mechanics will carry it and no farther. Time's cure for such an error is twofold: first, to leave it to work itself out and so refute itself; and secondly, to confront it with facts to which it will not apply. It was just such a conjuncture that made Bacon's denunciation of scholastic science effective. Perhaps some of you may live to see a second intellectual reformation in which the mechanical ideal of modern science will be proved in its turn to be defective and chimerical. At any rate, we have noted much that is ominous. Rigorously carried out as a theory of the real world, that ideal lands us in nihilism: all changes are motions, for motions are the only changes we can understand, and so what moves, to be understood, must itself be motion. Again, regarded as a descriptive or symbolic scheme, it proves to be only approximate and to become involved in interminable complications in the attempt to be exact. Just when scientific men, who are neither mathematicians nor physicists, Du Bois-Reymond and Huxley, for instance, are preaching "the advancing tide of matter and the tightening grasp of law," we find professed physicists renouncing their allegiance to this ancient idol. It is remarkable, too, that a change of a precisely opposite kind is going on in the more concrete sciences, which were formerly distinguished, as natural *history*, from physics, to which was reserved the title of natural *science*. Boltzmann refers to this: thus, he says: "What were formerly called the descriptive natural sciences triumphed, when Darwin's hypothesis made it possible, not only to describe the various living forms and phenomena, but also to explain them. Strangely enough, physics made almost exactly at the same time a turn in the opposite direction,"[1] *i.e.* as I understand, abandoning the attempt to be explanatory and contenting itself with being descriptive.

But returning now to the new theory of Energy. One important point for us to take account of — let me observe once more — is that this doctrine only entitles the physicist to assert the quantitative equivalence of phenomena that are qualitatively diverse: so much energy in the form of heat is equivalent to so much energy in the form of mechanical work; or again, so much thermal or mechanical energy has its equivalent in radiant energy or in energy of electric field. But it is going altogether beyond the facts to assume that all these forms are at bottom the same, *i.e.* mechanical or kinetic. The endeavour to reduce them to one is of course legitimate and in the interests of simplification. It is, however, pure hypothesis; there is no necessity about it; and, moreover, it is a hypothesis, as we have seen, round which, in spite of all that it has accomplished, difficulties seem steadily to thicken.

There is still another point that we must not overlook, — not only are the several forms of energy qualitatively distinct, but we have, I take it, no means of knowing that all these forms have been ascertained. We have no means of ear-marking a portion of energy; and it is not necessary to know all the transformations

[1] *Methods of Theoretical Physics*, Phil. Mag. 1893, vol. xxxvi, p. 40.

and transferences that may intervene in the course of a reversible cycle before it can be said that, whatever changes energy undergoes, it is never destroyed. Indeed it would, I believe, be substantially true to say that it was by assuming the conservation of energy, while still mistaken as to the nature of heat, that Carnot laid the foundation of thermodynamics. A strict *quid pro quo* is the one thing essential. The Bank of England issues notes equivalent in value to the gold in its cellars, and pays the gold out again to whoever presents the notes, and is so far unconcerned as to all the transactions that have intervened. Whether these transactions were many or few, domestic or foreign, industrial or financial — is of no account. So here: our ignorance of one or many possible transformations does not affect the main doctrine, provided we never find a transformation in which energy appears or disappears, unaccounted for.

But it is obvious that this possibility of unknown forms of energy coupled with the probability that the known forms are not all mechanical, suggests many new vistas, for which it behoves us to keep an open mind. I shall hope to recur to this briefly in dealing with psychophysics. For the present I think we are entitled as spectators of the march of science to say at least this much: Mechanics is no longer, at the end of the nineteenth century, what she was at the beginning, when the author of the *Mécanique céleste* proposed that "jubilant toast" to her that has served as our text. Absolute supremacy is hers no more; at best she is but *prima inter pares*, and even this, not because of the paramount value of the real knowledge she can bestow, but solely for her abstract purity of form. Should the science of energetics be destined to grow in importance at her expense, such an event would be by no means without a precedent. Think of the simplicity of the old Ionian and other pre-Socratic philosophies. Without a vestige of that knowledge that looms so large and imposing in the present concrete sciences, they set up their several ἄρχαι or first principles, water, air, fire, and so on; which now, so far from standing out as the obvious Alpha and Omega of all things, are simply lost in the multitude of particulars, quite on a par with them. And so in the history of science, do we see *axiomata media*, or middle principles, continually dwarfing and overtopping what had appeared as the veritable summits of knowledge in earlier days—such supposed summits constituting, by the way, the philosophy rather than the science of the time. And the remark is relevant, for mechanics, as I have had occasion to say before, has hardly yet ceased to count as natural philosophy, and even carries back its claims to those early times just referred to, when Democritus and Leucippus first broached the atomic theory. Its long supremacy is due largely no doubt to that vividness and mathematical accuracy with which the imagination can follow geometrical constructions. We are familiar with the influence of this fact, direct and indirect, on the minds of Plato, Descartes, Spinoza, and Kant. Had the inadequacy of the old atomism been realised earlier, the sway of the strictly mechanical theory would have been briefer. But it was only as physics and chemistry grew that these defects of the theory of "hard, massy particles" disclosed themselves in the course of attempts to resolve physical and chemical phenomena into mechanical processes between such particles. The result, as we have seen, has been to justify Lagrange's contention that mechanics is essentially a branch of pure mathematics, and as such subservient to, not dominant over, the concrete physical sciences. These meanwhile have a new ground of unity in the doctrine of energy. The only way to a supreme generalisation concerning physical things seems to lie through this; but it is altogether premature to suppose that that generalisation will be found to consist of such a world-formula as Laplace in his enthusiasm ventured to predict.

I have said much of this projected science of energetics, but nothing as yet of its main principle, the so-called Conservation of Energy. What does this mean? Methodologically, in other words, as a formal and regulative principle, it means much; really it means very little. Those who imagine that it furnishes any basis for statements concerning the past, present, or future of the universe, as a whole, are assuredly mistaken. And there are many such. We had an instance, for example, in the passage from Du Bois-Reymond's famous Leipzig address, which I quoted in the second lecture. Referring to Laplace's imaginary intelligence, Du Bois-Reymond represents him as calculating at what moment the universe will lapse into icy chillness, its energy, though conserved, being, in accordance with the second law of thermodynamics, entirely degraded to the unavailable form of heat at one temperature. To say nothing of the impropriety of treating the doctrine of the dissipation of energy as comparable in validity with the principle of the conservation of energy, the gratuitous assumption is here made that the quantity of energy in the universe is finite. If it should be infinite — and why should it not be? — then even Laplace's superhuman intelligence would be effectually nonplussed. But all statements concerning concrete quantity, and energy is such, imply measurement. There is but that one way of answering the question: How much? It cannot be answered *a priori* or by mere mathematics. To those who are fond of the 'high priori road' I will suggest the following consideration: If the energy of the world is a finite quantity and the second law of thermodynamics valid, how is it that the said degradation and consequent icy stillness are not the fact? On these assumptions the universe can only last a finite time, and the ratio of finite time to infinite duration is strictly infinitesimal. The chances then are infinity to one in favour of the universe being at any given moment 'played out.'

But now I will venture to say that not only does the principle of the conservation of energy tell us nothing about the quantity of energy in the universe as a whole, but that it does not even allow us to say that such quantity is an amount eternally fixed. I am quite aware that Mr. Spencer may here interpose with his caveat against "pseudo-thinking," and remind us of "the experimentally established induction" that energy is indestructible. As to the first — we shall come to the second presently — I am content to make again the reply made when we were discussing the conservation of mass. Reality and substantiality are not identical; if energy be verily and absolutely substantial, it must no doubt be verily and absolutely permanent, neither generated nor liable to decay. But it is obvious that we cannot by observation or measurement show that this is actually the case, nor can we by *a priori* reasoning prove that it necessarily must be. It would be safe to go further, and to say that if energy were verily and absolutely the substance of things, it could not be *measured* at all. To what is absolutely substance the notion of unity and totality will apply, but these are not metrical notions. The scientific meaning of the statement, "the energy of the universe is constant," then, is not what at first blush it seems to be and is often mistaken to be. Apparently an absolute statement, it is really a relative one, and only valid as such. Apparently a statement of fact, it is really only a postulate. As with the conservation of mass, which — as we have seen — it may turn out to include, so with the conservation of energy; there are the same two grounds for making it, but neither will suffice to place it beyond question. First, it is borne out by experience, so far as we know; and secondly, it seems the simplest and best working hypothesis. As to its relativity, this it shares in common with every other empirical statement: all such tell us nothing but the ratio between the quantity measured and the quantity of the unit or standard employed in measuring. If both these quantities were to vary in the same proportion, their ratio, of course, would remain unaffected; hence it can afford no evidence of such variation. We assume, however, that our standard is fixed, or what comes to the same thing *for metrical purposes*—that, if there is any variation, it is a uniform variation throughout the universe. This is all that constancy means. But a principle that will allow of such an interpretation cannot be one relating to substance.

Regarded as a postulate the conservation of energy appears under a somewhat different aspect, and one of especial interest to us. I greatly regret that there is not time enough left to deal with it more fully. It is allowed that as an experimental generalisation the conservation of energy can only claim to be probable; on what ground then is it put forward as a fundamental principle? Helmholtz, also Thomson and Tait, found on "the axiom that the Perpetual Motion is impossible." Mayer, a genius to whom the world has yet to do justice, and even Joule, are more 'metaphysical.' Mayer falls back on the formula, *Causa æquat effectum;* and Joule declares it "manifestly absurd to suppose that the powers with which God has endowed matter can be destroyed."[1] It is clear, then, that not only are we not here in the region of experimental proof, but that no direct proof of any kind is offered us. The use of terms such as 'impossible' and 'absurd' shew plainly that any proof there is, is indirect—a sure sign that, if we are dealing with a truth at all, it is one that is self-evident. And yet it was not till the year 1775 that the French Academy of Sciences, with Lagrange and Laplace at their elbow, were so far convinced that

[1] Cf. Mach, *Popular Scientific Lectures*, p. 246.

the perpetual motion was impossible, as to decline for the future to receive any pretended demonstration of such a machine. Moreover, as Mach[1] has pointed out, the principle of virtual velocities, on which Lagrange's whole *Méchanique analytique* rests, really presupposes this axiom; yet Lagrange himself was not clearly aware of it, though sensible of the insufficiency of his proof as it stood — an insufficiency that led Poinsot to remark, that Lagrange had only lifted the clouds from the course of mechanics, because he had allowed them to gather at the very origin of that science. But after all the impossibility of perpetual motion only covers half the ground; friction and strain are absent from ideal mechanisms, so that the question what goes with apparently wasted energy does not arise. It was the study of actual machines, with which Lagrange never troubled himself, that brought this side to the fore; and it is this, the converse of the first axiom, that Joule is attempting vaguely to formulate when he says it is absurd to suppose that material powers can be destroyed. The remark is noteworthy, for it is customary to extol Joule as a sound experimentalist and to depreciate Mayer as a metaphysical dreamer. But there is little doubt that both men first conceived the general truth, and then set about — the one by experiments, the other by computations from ascertained physical constants — to verify what they had thus conceived. Mayer in one of his letters, quoted by Mach, says expressly: "Engaged during a sea voyage almost exclusively with the study of physiology, I discovered the new theory for the sufficient reason that I *vividly*

[1] Mach, *Lectures*, pp. 152 f.

felt the need of it."[1] But Mayer's statements are the more comprehensive inasmuch as he refers to both the creation and the annihilation of energy as impossible assumptions, summing up both in the one formula, *Causa æquat effectum.* To be sure this as it stands is too vague and perhaps too general to be impressive. More definite and workable formulations have been devised since. But the point is that, in however imperfect a form, Mayer's statement of the principle embodies all that is axiomatic in the conservation of energy, and that at bottom is none other than the principle of sufficient reason which you will remember Laplace too postulated. More precisely — since in dealing with energy, we are dealing with procession, with changes — the axiom implied is the principle of causality. These two principles of sufficient reason and causality may occupy us at some length later on. But I will anticipate to the extent of mentioning some points that will help us to round off this portion of these lectures, and bring it not merely to an end, but to some sort of conclusion.

Looked at broadly, if you will philosophically, these principles of sufficient reason and causality are part of the postulate that everything shall be intelligible and the whole of things rational. This is the *faith* of science; on this point all are agreed. Even Hume and Kant are here at one; both allow that such principles do not derive their validity from experience, though they differ widely as to what this validity is worth. The principle of causality is not a logical or a mathematical, but a real princi-

[1] Mach, p. 184.

ple; in the principle of the conservation of energy we have it in a quantitative form applied to physical changes. So we may see by the way how Lagrange as the representative of abstract mechanics failed to reach it, while Mayer, bent on rendering concrete physical facts intelligible, "vividly felt the need of it."

But though a real principle, the conservation of energy only renders the quantitative relations of physical processes intelligible. What about the qualitative relations between which it only determines quantitative equivalences? Have we not an equal right to postulate intelligibility here too? It is here that the psychical as distinct from the physical comes in. Action initiated by feeling is now the fundamental fact. True, we still have quantitative distinctions of a sort; that is, we have a scale of values or worth, degrees of pleasure and pain, degrees of beauty and ugliness, degrees of merit and demerit. But qualitative differences not amenable to mathematical treatment underlie them all. Motives, then, are of the nature of causes; and conduct falls within the range of the principle of sufficient reason; although in the last resort conduct carries us back to a sentient being with its pronouncement, *Sic volo, sic jubeo, stet pro ratione voluntas.* Let me recall your attention to two points in the famous pæan of Laplace: (1) his acceptance of the principle of sufficient reason as fundamental; and (2) his assumption that his imaginary intelligence "shall be acquainted with all the forces [let us say, with all the causes] by which nature is animated." If pleasures and pains can be sufficient reasons, they too must be reckoned among the causes that animate nature, or at least among the causes that determine events. Laplace, no doubt, was careful to rule out free will; but that is not enough. Quite apart from the difficulties of that venerable problem, motives remain as a class of causes not yet admitting of mathematical treatment, still less of mechanical interpretation. *De gustibus non est disputandum* here passes from a mere maxim almost into a metaphysical principle. In other words, wherever there is feeling and preference there is something unique. Now, either this uniqueness appears in the physical world or it does not. The admission that it does will make it very difficult to stop short of regarding all the beings that compose the world — so far as 'being' implies any sort of unity or individuality — as feeling-agents, monads, or 'mind-stuff.' Now, though such an admission might still leave room for an omniscient Deity, it would, it seems to me, make an end of the Laplacean physicist. Kant saw this very clearly; unhappily Clifford and other physicists, who have a predilection for 'mind-stuff,' do not seem to see it. "Life," says Kant, "means the capacity to act or change according to an internal principle. But we know of no internal activity whatever but *thought*, with what depends upon it, feeling of pleasure or pain and desire or will. But matter is lifeless, for on the law of inertia (next to that of the permanence of substance) the possibility of physics proper entirely depends. The opposite of this, and therefore the death of all natural philosophy, would be Hylozoism." [1] By the death of all natural philosophy, however, Kant means only that the

[1] *Metaphysische Anfangsgründe der Naturwissenschaft*, Hartenstein's edition, vol. iv, p. 439.

mechanical theory would lose its supremacy; and that in 1786 was a thing not to be thought of. Just a century later, in 1886, we have a distinguished organic chemist, Bunge, declaring "*So treibt uns der Mechanismus der Gegenwart dem Vitalismus der Zukunft mit Sicherheit entgegen*";[1] the mechanical theories of the present are urging us surely onwards to the vitalistic theory of the future. It is mainly the tyranny of imagination that is in the way. Picture the position of Galileo, to whom the mechanical theory is primarily due, and it will be easier to believe in the Galileo that is to be.

Meanwhile, the view holds its ground that the uniqueness of feeling agents does *not* affect the physical world. To prevent "the death of all natural philosophy," it is maintained that the psychical never affects the physical sphere, the two being pronounced utterly distinct, disparate, and, so to say, incommensurable. But what if there are not two spheres; and if only one, what if the psychical is that one? However, assuming the dualism now prevalent among scientific men, according to which life and mind are merely impotent concomitants of the physical, epiphenomenal as the latest phrase is — it is difficult to see that the Laplacean physicist will be any better able than before to peer into past or future history. Grant that he knows all the changes of any brain he may select as accurately as he knows the phases of the moon. Yet he only knows them in the same way, *i.e.* as material events. As such, they afford, *ex hypothesi*, *no* clue to their mental concomitants; nay, it is of the very essence of the hypothesis that they should afford no clue.

[1] *Vitalismus und Mechanismus*, ein Vortrag., p. 20.

Such dualism, it has been said, means chopping the world in two with a hatchet. It is indeed a murderous stroke, and leaves us with two dead and impotent halves in place of the living whole. Or worse, it gives us two sets of abstractions in place of one reality. This comes out in an odd way when we compare the deliverances of many of our physiological teachers with those of foremost physicists of the Kirchhoff school. Huxley, for example, thus winds up his article on Conscious Automatism: "If these positions are well based, it follows that our mental conditions are simply the symbols in consciousness of the changes which take place automatically in the organism; and that, to take an extreme illustration, the feeling we call volition is not the cause of the voluntary act, but the symbol of the state of the brain which is the immediate cause of that act." There seems then no escape from the conclusion that the whole world is symbols. Attractions, affinities, undulations, molecules, atoms, ether, are to be regarded primarily as "mere helps or expedients to facilitate our viewing things," not as the veritable realities: so Kirchhoff or Mach. But on the other hand the 'perceptual realities,' which those physicists are content to recognise, are simply shadows and symbols: so the physiologists.

Have we no means of deciding the question at issue: Which is the real and which is the symbolic? If the question is fairly faced, it seems to me the answer is extremely easy. Roundly stated, the real is always concrete, the symbolic is always abstract. The real implies individuality more or less; the symbolic is always a logical universal. Within the range of our experience the real implies always a history, that is, places and dates, converse with a concrete environment. The symbol is the creature of logic. If temporal and spatial relations enter into its definition or description, they are time and space coördinates with no vestige of chronology or topography about them. Now, tried by this standard, it is a glaring absurdity to call Cæsar's resolve to cross the Rubicon or Luther's to enter Worms the symbol of the dance of molecules in their brains. Yet to this pass Huxley brings himself. As I have tried to shew, and as I believe, the very advance of physics is proving the most effectual cure for this ignorant faith in matter and motion as the inmost substance rather than the most abstract symbols of the sum of existence.

And what, it may be asked, do I mean to argue from this? Simply that in our speculation about the universe we should never let go the concrete that we envisage. As long as we keep to that we find no two things absolutely alike, no two events absolutely the same. Intellectually to compass the wealth of particulars we are driven to generalise and symbolise, to employ the instrumentality of identity and uniformity among substances and causes, when the full fact is development and progress. It is far truer to say the universe is a life, than to say it is a mechanism, even such a mechanism as Goethe describes in verses that German men of science are fond of quoting, where the Spirit of the Earth "weaves at the rattling loom of the years the garment of Life which the Godhead wears." We can never get to God through a mere mechanism. I should not like to pin my faith to Leibnitz, but of all the dogmatic philosophies his seems to me — in one feature at any rate — by far the best. With him, then, I would argue that absolute passivity or inertness is not a reality, but a limit. I would not say that the atoms of our present physicists are monads, for it is still an open question if they are anything. But to whatever is entitled to be called "one of the beings composing the world," — Laplace's phrase, you will remember, — I would ascribe enough initiative and individuality to put his famed Intelligence to confusion.

PART II

THEORY OF EVOLUTION

THEORY OF EVOLUTION

LECTURE VII

MECHANICAL EVOLUTION

1 Mechanical evolution, *the process by which the mass and energy of the universe have passed from some assumed primeval state to that distribution which they now present. Mr. Herbert Spencer the best accredited exponent of this doctrine.*

He regards the universe as a single object, which is alternately evolved and dissolved. But the universe cannot be so regarded; and, if it could, Mr. Spencer's mechanical principles forbid such alternation. He ignores 'dissipation of energy,' and confuses energy with work. The thermodynamic zero. A finite universe must have time limits.

But is the universe finite? The Kantian antinomies and their solution. The notion of evolution not applicable to 'the totality of things.'

2 The doctrine of the dissipation of energy and questions of reversibility. *Limitations introduced by Lord Kelvin, Helmholtz and Maxwell. Two alternatives thus appear equally compatible with Mr. Spencer's 'fundamental truth.'*—(a) *evolution without guidance, and* (b) *evolution with guidance. To account for the visible universe according to* (a) *requires a definite 'primitive collocation.' This Mr. Spencer rejects; for him then the cosmos can be but a chance hit among many misses, a mere speck of order in a general chaos. In expecting more from his mechanical principles he is guilty of the fallacy of confounding* (a) *with* (b).

In resuming our discussion after so long an interval it may be well briefly to restate what it is that we have set out to discuss. Naturalism we have taken to designate the doctrine that separates Nature from God, subordinates Spirit to Matter, and sets up unchangeable law as supreme. It means, to quote again the words of Huxley, "the extension of the province of what we call matter and causation and the concomitant . . . banishment from all regions of human thought of what we call spirit and spontaneity . . . [till] the realm of matter and law is coextensive with knowledge, with feeling, with action."[1] This naturalistic philosophy consists in the union of three fundamental theories: (1) the theory that nature is ultimately resolvable into a single vast mechanism; (2) the theory of evolution as the working of this mechanism; and (3) the theory of psychophysical parallelism or conscious automatism, according to which theory mental phenomena occasionally accompany but never determine the movements and interactions of the material world. With the first of these we have already dealt, and we now come to the second, in which it is applied.

Yet evolution, as commonly understood, is as far as possible from suggesting mechanism. By evolution or development was meant primarily the gradual unfolding of a living germ from its embryonic beginning to its final and mature form. This adult form, again, was not regarded as merely the end actually reached through the successive stages of growth, but as the end aimed at and attained through the presence of some archetypal idea, entelechy, or soul, shaping the plastic material and directing the process of growth. Evolution, in short, implied ideal ends controlling physical means;

[1] Cf. Lecture I, p. 17.

in a word, was teleological. In this sense *mechanical* evolution or development becomes a contradiction in terms. Nevertheless we shall find that the category of End, equally with the categories of Substance and Cause, is nowadays outside the pale of natural science. The term 'evolution,' though retained, is retained merely to denote the process by which the mass and energy of the universe have passed from some assumed primeval state to that distribution which they have at present. Also it is implied that the process will last till some ultimate distribution is reached, whereupon a counter-process of dissolution will begin. Let us now turn to Mr. Herbert Spencer, the best accredited exponent of this doctrine, for details.

"An entire history of anything," Mr. Spencer tells us, "must include its appearance out of the imperceptible and its disappearance into the imperceptible. Be it a single object or a whole universe, any account which begins with it in a concrete form, or leaves off with it in a concrete form, is incomplete." "The sayings and doings of daily life," he continues, "imply more or less such knowledge of states which have gone before and of states which will come after. . . . This general information which all men gain concerning the past and future careers of surrounding things, Science has extended, and continues increasingly to extend. To the biography of the individual man, it adds an intra-uterine biography beginning with him as a microscopic germ; and it follows out his ultimate changes until it finds his body resolved into the gaseous products of decomposition." So as to the clothes he wears—"not stopping short at the sheep's back and the caterpillar's cocoon, it identifies in wool and silk the nitrogenous matters absorbed by the sheep and the caterpillar from plants." So also as to "the wood from which furniture is made, [this] it again traces back to the vegetal assimilation of gases from the air and of certain minerals from the soil. And inquiring whence came the stratum of stone that was quarried to build the house, it finds that this was once a loose sediment deposited in an estuary or on the sea-bottom." In these and such like instances Mr. Spencer sees the formula of evolution and dissolution foreshadowed. To quote again his own words: "In recognising the fact that Science, tracing back the genealogies of various objects, finds their components were once in diffused states, and pursuing their histories forwards, finds diffused states will be again assumed by them, we have recognised the fact that the formula must be one comprehending the two opposite processes of concentration and diffusion. . . . The change from a diffused, imperceptible state, to a concentrated, perceptible state, is an integration of matter and concomitant dissipation of motion; and the change from a concentrated, perceptible state, to a diffused, imperceptible state, is an absorption of motion and concomitant disintegration of matter."[1]

Now, there is one obvious yet serious objection to this theory. It proposes to treat the universe, in fact requires us to treat the universe, as we treat a single object. Every single object is first evolved and then dissolved; it emerges from the imperceptible and into

[1] *First Principles*, pp. 279–281.

the imperceptible it disappears again. And so of the universe: "Any account which begins with it in a concrete form or leaves off with it in a concrete form," Mr. Spencer tells us, "is incomplete." Surely we have here a case of what logicians call the fallacy of composition; what is predicable of the parts severally is predicated of the whole collectively. It reminds us forcibly of Locke's "poor Indian philosopher, who imagined that the earth always wanted something to bear it up." The stability of everything on the earth was manifestly due to a support, therefore the stability of the solid earth itself seemed explicable in no other manner. So the poor Indian; and similarly Mr. Herbert Spencer. As science deals with any visible, tangible thing, so the "synthetic philosophy" will deal with the totality of things. Let us take as a simple instance of the first, the familiar case suggested by Mr. Spencer himself, that of a cloud appearing when vapour drifts over a cold mountain top, and again disappearing when it moves away into the warmer air. The cloud emerges from the imperceptible as heat is dissipated and the vapour condensed, and the cloud is dissolved again as heat is absorbed and the watery particles evaporate. How shall we apply this conception or anything like it to the universe? The stronghold of Mr. Spencer's argument is the nebular hypothesis. A nebula, no doubt, is an object among other objects, though a most sublime and stupendous one. It presupposes colliding stars or meteoric swarms, whose material constituents are dissipated by the heat which their collision has produced; but then these colliding masses in their turn imply still earlier nebulæ, whose materials concentrated as their heat diffused. So the cloud presupposed vapours that had previously condensed; and the vapour, cloud or water that had previously evaporated. And much as clouds dissolve in one place and form in another, and are to be found at any time in all possible stages of evolution and dissolution; so with sidereal systems and nebulæ. The telescope and spectroscope tell of stars and nebulæ in every phase of advance or decline to be found in every quarter of the heavens. To ask which was first, solid masses or nebulous haze, is much like asking which was first the hen or the egg, and like that famous problem, may lead us to conclude, — neither the one nor the other. Meanwhile, it does not surprise us to learn that, though Mr. Spencer is quite sure that the universe began as imperceptible mist, others, like the late Dr. Croll, who have incomparably more right to an opinion on the question, prefer to think that there was an earlier or præ-nebular stage of the universe; during which large, cold masses of protyle or primal matter were moving through space in all directions with excessive velocities.[1] Such an hypothesis, whether otherwise admissible or not, at least recognises a problem with which Mr. Spencer scarcely attempts to deal — I mean the evolution of the chemical elements. It thus suffices to convict Mr. Spencer's work of a certain incompleteness. For surely to begin with some seventy distinct forms of matter with very various and definite properties is not to begin at the beginning, however much we may imagine them to be diffused. We must return to

[1] Cf. Croll, *Stellar Evolution*, pp. 3, 109.

this question of qualitative diversity presently. But the prior question I am anxious to put as pointedly as possible is this: On what grounds is it assumed that the universe was ever evolved at all? A given man, a given nation, a given continent, a given sidereal system, as particular objects, have their several finite histories of birth and death, upheaval and subsidence, fiery mist and cold, lifeless, consolidation. But growth and decay, rise and decline, elevation and degradation, evolution and dissolution, are everywhere contemporaneous. We have but to extend our range to find a permanent totality made up of transient individuals in every stage of change. But so enlarging our horizon we are not warranted in saying, as Mr. Spencer does, "there is an *alternation* of Evolution and Dissolution in the totality of things."[1] Of the totality of things we have no experience. But now what we do find, so far as experience and observation will carry us, is that, be it great or small — once an object has disappeared into the imperceptible, once it is dissolved in Mr. Spencer's sense, that object never reappears. We do not find dead men alive again, effete civilisations rejuvenated, denuded continents again restored, or worn-out stars rekindled as of yore. *If* there were any justification for the phrase "visible universe" and *if* we could conceivably represent the totality of things as a single concrete *object*, — both which suppositions I deny, — then by all analogy and experience 'alternate eras of Evolution and Dissolution' would be physically impossible. So surely as 'the appearance out of the imperceptible' was the beginning, so surely would

[1] *First Principles*, p. 551.

'the disappearance into the imperceptible' be the end. As, according to Mr. Spencer's own description, the entire history of anything, be it a single object or the whole universe, lies completely within such limits, it is a manifest contradiction to turn round and say: After all the end is not the end and the beginning is not the beginning, and what we have called an entire and complete account of the totality of things is only one wave in an endless rhythm. It is true, of course, that the history of many concrete objects is marked by periodic phases; but never by dissolution and reëvolution, *i.e.* by the disappearance of the concrete individual followed by the reappearance of that individual—in short, by what is tantamount within the scope of such terms as visible, tangible, concrete, and perceptible—to as complete a breach of individuality as we should have in annihilation and re-creation. It is also true, as we have already noted, that *within* a given totality, one individual may succeed another, but so far that totality—the universe of discourse, so to say—remains permanent. "One generation passeth away, and another generation cometh: but the earth abideth for ever."[1]

Moreover, on the physical assumption from which Mr. Spencer sets out, viz. that the mass of the universe and the energy of the universe are fixed in quantity—which ought to mean are finite in quantity—there can be no such alternations as he supposes. Certainly not if we are to accept the second law of thermodynamics, the law, that is, of the dissipation of energy, along with the first law, that of the conservation of energy. But of this

[1] Eccl. i. 4.

second law, commonly accepted though it is by physicists at the present day, Mr. Spencer seems to take no account. Apparently, too, Mr. Spencer confuses energy or the *capacity of doing* work with work actually *done*, and imagines that so long as the quantity of energy persists, it must be manifest in perpetual changes of equivalent amount. But this in any case is not a necessary consequence of the conservation of energy, and if the dissipation of energy be true, it is an impossible consequence. For it is not on the bare persistence of energy, but on the transference and transformation of energy that physical changes depend. But energy, whatever be its form, is only transferable from places of higher 'intensity' to places of lower intensity, to use a convenient term. So we find heavy bodies tend to fall, hot bodies to cool, and so forth. Thus the amount of energy *available for work* of the total energy possessed by two bodies is a function of this difference of level or intensity, and is *nil* when this difference is *nil*, whatever the total energy may be. Generally speaking, energy is not transferred without an exact equivalent of work being done; but to this rule thermal energy is an exception. And it is here that the so-called waste or dissipation of available energy comes in. Putting it quite popularly, in the partnership of energies, heat is the one squanderer, and may scatter without producing. Whenever energy passes into this form, some of it is always, and all of it is sometimes, lost for purposes of work. As Mach puts it, "heat is only partially transformed into work, but frequently work is wholly transformed into heat. Hence a tendency exists towards a diminution of the *mechanical* energy and towards an increase of the *thermal* energy of the world."[1] In other words, though the energy of the world remains constant, the unavailable energy or entropy, as it is called, tends towards a maximum. There is still a peculiarity of heat to be mentioned that will make the significance of the thermal degradation of energy clearer — I refer to Lord Kelvin's definition of an absolute zero of temperature. If — whatever were the temperature of a body — we could always imagine another body with a temperature still lower, just as whatever be the position of a body we can always imagine another at a distance from it towards which it can gravitate, then, so far as in this way differences of temperature would always be possible, the transformation of heat into work might always be possible. But if there be, as is supposed, a thermodynamic zero, there is an end to such a possibility; beyond that zero temperature cannot fall. And so while all transformations of energy lead directly or indirectly to transformation into heat, from that transformation there is no complete return, and therefore finally no return at all. This then is the conclusion to which Mr. Spencer's premisses lead. Two eminent physicists who accept those premisses may be cited at this point: "It is absolutely certain," they say, "that life, so far as it is physical, depends essentially upon transformations of energy; it is also absolutely certain that age after age the possibility of such transformations is becoming less and less; and, so far as we yet know, the final state of the present universe must be an aggregation (into one mass) of all the matter it contains, *i.e.* the potential

[1] *Popular Scientific Lectures*, Eng. trans., p. 175.

energy gone, and a practically useless state of kinetic energy, *i. e.* uniform temperature throughout that mass. . . . The present visible universe began in time and will in time come to an end."[1]

To this conclusion we are surely led from such premisses. But again I ask what warrant is there for the premisses? Our experience certainly does not embrace the totality of things, is, in fact, ridiculously far from it. We have no evidence of definite space or time limits; quite the contrary. Every advance of knowledge only opens up new vistas into a remoter past and discloses further depths of immensity teeming with worlds. The physical principles of the conservation of mass and energy are, as I have already urged, essentially formal and regulative; they do but formulate the common postulate of all science — the uniformity and continuity of nature as presupposed in all physical measurements. They do *not* justify us in assuming, what we certainly cannot prove, that the universe as a whole is measurable and therefore finite. And when we pass to more purely *a priori* considerations, the case against a universe with fixed and finite limits is equally strong. It is needless to attempt even the most cursory discussion of the antinomies as to the finitude or infinitude of the universe in respect of time, space, divisibility, or mass, that have constituted the chief cosmological problems of philosophy, notably since the time of Kant. They have only justified in the main Kant's own solution. We cannot say that the phenomenal universe is infinite in any of these aspects,

[1] *The Unseen Universe*, § 115.

but just as little can we say it is finite. Since Kant's day, more cogent arguments both for the theses and for the antitheses of the cosmological problem have been advanced. None of these invalidate the claims of reason to regard the universe as a systematic whole, but they set in a stronger light than ever the impossibility of treating it as an arithmetical sum. "Say that the universe is limited," says Kant, "and it is too small for your concept; you have a perfect right to ask what determines that limit: but say that it is unlimited, and it is too large for every possible empirical concept." The reason of this is plain. In the empirical regress, to which the understanding, that is science, is entirely confined, "no experience of an absolute limit, that is, of any condition as such, which empirically is absolutely unconditional, can exist." On the other hand, this regress from any given phenomenon as conditioned to another as its condition, though not truly infinite, is never suspended yet never completed; in other words, such regress must proceed *in indefinitum*.

But what Mr. Spencer calls a single object, must surely have an assignable beginning and end in time and assignable bounds in space; it is precisely through such time and space marks that the notion of singleness or identity becomes applicable. Those marks, however, are not given by empty time or space, but by other objects relatively defined in the same fashion. The universe, then, we may safely say, not only *is* not, but never can be, a single object in this wise; and Mr. Spencer's attempt to treat it after the fashion of an evolving nebula, evinces an unexpected paucity of imagination and is philosophically unsound. Experience provides us with instances of evolution and dissolution on the most varied scales, from the grass of the field or the cedars of Lebanon to the solar system or the Milky Way. But of a single supreme evolution embracing them all we have no title to speak: not even to assume that it is, much less to say what it is; least of all to affirm confidently that it can be embraced in such a meaningless formula as the integration of matter and the dissipation of motion—doubly meaningless unless a partial system, such as a nebula, is concerned, and even then assuming the greater portion of molecular physics without explanation. We have no evidence to shew that what we miscall the 'visible universe' is coming to an end, for we have no evidence to shew that it is finite. If we had such evidence, we should probably then and there conclude that we were dealing with but a part of the true universe and not with the totality of things. Again there is no physical evidence to compel the application to this absolute totality of such conceptions as increase and decrease, ebb and flow, development and decay; no warrant for attributing to the universe a destined perfection, that if future must either be attained and past; or approached but never completely attained at all. The former, if, as Mr. Spencer supposes, the mass and energy of the universe are finite and fundamental; the latter, if, being still the fundamental factors, the mass and energy are mathematically infinite in amount. Whether the world be absolutely perfect, or merely the best of all possible worlds, or indeed the worst world possible, as actual, it is—so far as we can judge from its physical constitution — just what it always has been, the permanent theatre of perpetual changes.

At any rate such a conception is less conjectural and more adequate than Mr. Spencer's ridiculous comparison of the universe to a spinning top that begins by 'wabbling,' passes into a state of steady motion or *equilibrium mobile*, and finally comes to rest. Referring to this second phase as one of perfect moving equilibrium, he finds in it "a warrant for the belief that evolution can end only in the establishment of the greatest perfection and the most complete happiness." Let us not pause now to ask what sort of perfection and happiness that must be which depends on and necessarily follows from such physical equilibration: let us note only that, whatever it be, it is after all, according to Mr. Spencer, neither final nor established. It is but the "penultimate stage," as indeed he calls it, and gives place, as he tells us, to "Dissolution, which inevitably, at some time or other, undoes what Evolution has done."[1] And again I say that the absurdity to which Mr. Spencer betakes himself does not suffice to put a better face on his doctrine — the absurdity, I mean, of supposing that, though there cannot be two universes in space, there may be any number in time. Beyond the final *quietus* of cosmical equilibration the doctrine of energy, in which Mr. Spencer puts his trust, affords no hope of a new evolution. The dead bones, the black ashes, may or may not live and glow again, but if they do it will not be from the mere 'persistence of

[1] *First Principles*, p. 550.

force' that the quickening burst will come. Why, if the thing is so obvious, not to say necessary, is it never elucidated by 'the familiar example' of the spinning top? No doubt *two* consolidated sidereal systems may diffuse again after a collision, but how is *one* to do this? And what can well be less suggestive of recurring cycles than universal concentration of mass and uniformity of temperature on the one hand and indefinite diffusion of mass and diversity of temperature on the other? It must be allowed that in so far as Mr. Spencer is personally concerned, the doctrine of the dissipation of energy was scarcely in the air when his First Principles were stereotyped. Meanwhile, for us at any rate, that doctrine seems to put an end to the alternate eras of evolution and dissolution which Mr. Spencer has vainly striven to derive from the doctrine of conservation. On the whole then we may for the present reasonably demur to Mr. Spencer's attempt to bring the universe under a simple formula of evolution and dissolution, as if it were a single object emerging out of the imperceptible and dissolving into it again. Before proceeding to discuss his formula in more detail so as to ascertain its adequacy where evolution in some sense is admissible, let me ask your attention for a little longer to one or two reflections suggested by our inquiry thus far or by points incidentally raised in the course of it.

Among the last in particular is this doctrine of the dissipation of energy, which excludes such reversibility as Mr. Spencer supposes. Lord Kelvin, who was, I believe, the first to formulate this doctrine, has been frequently commended for the caution which led him to restrict the impossibility to cases in which the agency of inanimate matter is alone concerned. Thus Helmholtz, referring to this reversion in a review of Lord Kelvin's papers, says: "Such a reversion is a postulate beyond the power of human means to fulfil. We have no agency at our disposal by which to regulate the movement of atoms. Whether, however, in the extraordinarily fine structure of organic tissues a mechanism capable of doing it exists or not is a question not yet to be answered, and I deem it very wise on the part of Sir W. Thomson that he has limited all his theses respecting the necessity of increasing dissipation by restricting their validity to 'inanimate matter.'"[1] Dissipation of energy Lord Kelvin himself tells us, "follows in nature from the fortuitous concourse of atoms. The lost motivity is essentially not restorable otherwise than by an agency dealing with individual atoms; and the mode of dealing with the atoms to restore motivity is essentially a process of assortment, sending this way all of one kind or class, that way all of another kind or class."[2] Many here will remember a fine passage in Mill's *Political Economy* on the function of labour, in which he shews with impressive detail that in what is called the action of man upon nature it is "the properties of matter that do all the work, when once objects are put into the right position. This one operation of putting things into fit places for being acted upon by their own internal forces, and

[1] *Wissenschaftliche Abhandlungen*, Bd. iii, p. 594.

[2] *Properties of Matter*, p. 139.

by those residing in other natural objects, is all that man does, or can do, with matter. He only moves one thing to or from another:" all his vast command over natural forces immeasurably more powerful than himself "is obtained by arranging objects in those mixtures and combinations by which natural forces are generated, as when by putting a lighted match to fuel, and water into a boiler over it, he generates the expansive force of steam, which has been made so largely available for the attainment of human purposes."[1] Here then we have the materials and powers of nature, as they fortuitously occur, incapable of, and unavailable for, results, to which nevertheless they can be guided by intelligent assortment and arrangement. And in a precisely analogous way we can imagine finite intelligences disequalising temperature and undoing the natural diffusion of heat, or assorting atoms and undoing the natural conglomeration of matter, and so reversing the downward trend, and even disturbing the final quiescence, to which the dissipation of energy or 'cosmic equilibration,' to use Mr. Spencer's term, inevitably leads. The conception of such an intelligence we have in "the sorting demon of Maxwell," as Lord Kelvin has called it.

This brilliant idea was devised by Maxwell primarily to illustrate "the limitation of the second law of thermodynamics," to shew, that is, that this second law, the law of the degradation of energy is not like the first—the law of conservation—a fundamental, dynamical law; that,

[1] *Principles of Political Economy*, Bk. i, chap. i, § 2. Mill attributes this observation to his father, but even he was anticipated by Bacon (*Novum Organum*, vol. i, p. 4), and again by Playfair.

on the contrary, it is properly a statistical law and confined to our experience of secondary bodies consisting of an immense number of molecules, none of which are individually perceptible. And so he remarks: "This law is undoubtedly true as long as we can deal with bodies only in mass, and have no power of perceiving or handling the separate molecules of which they are made up. But if we conceive a being," — and here we are introduced to the 'sorting demon' — "whose faculties are so sharpened that he can follow every molecule in its course, such a being, whose attributes are still as essentially finite as our own, would be able to do what is at present impossible to us." To most of you I am sure the *modus operandi* of this possible but imaginary being is perfectly well known; still, to add to the clearness of our discussion, I will venture to quote the rest of Maxwell's paragraph. "For we have seen," he continues, "that the molecules in a vessel full of air at uniform temperature are moving with velocities by no means uniform, though the mean velocity of any great number of them, arbitrarily selected, is almost exactly uniform. Now let us suppose that such a vessel is divided into two portions, A and B, by a division in which there is a small hole, and that a being, who can see the individual molecules, opens and closes this hole, so as to allow only the swifter molecules to pass from A to B, and only the slower ones to pass from B to A. He will thus, without expenditure of work, raise the temperature of B and lower that of A, in contradiction to the second law of thermodynamics."[1]

[1] *Theory of Heat*, 1894, pp. 338 f.

Now, what I think we may fairly deduce from this piece of physical exposition is that conservation of energy at any rate, — and this is Mr. Spencer's one dynamical principle, — is compatible with either of two alternatives.[1] The first is that steady fall in the level of available energy, which finds expression in the second law of thermodynamics, technically given in the statements of Lord Kelvin and Clausius already quoted,[2] viz., that, though the energy of the universe remains constant, the entropy of the universe tends towards a maximum.[3] The second alternative is the process of assortment and guidance — without expenditure of work — by a selecting and directing intelligence, which process may, to an indefinite extent, reverse and overrule the dissipation of energy, that tendency merely to run down. For, granting the energy of a material system, however large, to remain constant, and granting change of direction without work to be always theoretically possible, we may infer that, until — after a lapse of time indefinitely great — a state of absolute dissipation is reached, it would be possible for intelligent beings, without infringing any dynamical principles, to inaugurate changes, and for an adequate intelligence to start that system anew on a fresh round of evolution. This is forcibly put in the paper of Helmholtz's, from which I have already quoted: "The ascertained laws of dynamics," he says,

[1] But our difficulties, no doubt, increase when we take into account other dynamical principles which Mr. Spencer neglects. Cf. below, Lecture XII.

[2] Cf. above, p. 194.

[3] Strictly speaking, we are not warranted in applying metrical motions to the universe. Cf. pp. 87, 171.

"yield the deduction that, if we were able suddenly to reverse the total movements of the total atoms of an isolated mechanical system, the whole system would of necessity retraverse all the states which up to that point of time it had passed through. Therewith, also, would all the heat, generated by friction, collision, conduction of electrical currents, etc., return into other forms of energy, and the energy, which had been dissipated, would be all recovered."[1] And I presume that an intelligence that could precisely reverse the directions could alter them as easily in other ways. But the point is that, apart from intelligent guidance and arrangement, no such recovery or alteration would be possible.

It will be quite worth while to compare these alternatives somewhat further. Though both are equally compatible with the persistence of energy, yet Mr. Spencer, as we have seen, admits only one, and ignores the fact that that one entirely precludes such alternations of evolution and dissolution as he assumes. According to that, which we may fairly call the mechanical view, evolution, or rather, as Mr. Spencer ought to say, a given era of evolution, begins at an initial extreme, characterized by him as an imperceptible state of absorption of motion and concomitant disintegration of matter; and ends with a final extreme, equally imperceptible, of integration of matter and concomitant dissipation of motion. In conciser and more intelligible language, the whole process ranges from an extreme with very large potential energy to an extreme in which all available energy is dissipated. The other

[1] *Wissenschaftliche Abhandlungen*, Bd. iii, p. 594.

alternative, which we may perhaps call the teleological view, neither requires an initial stage, such as Mr. Spencer's, in order that evolution may begin, nor is debarred by the dissipation of energy from all possibility of further change. Without postulating the creation of energy it recognises the direction of energy by intelligence. Under what circumstances and by what means such intelligent guidance is effected we need not now inquire; it is allowed to be possible, and for the present that is enough.

And now let us attend to the important difference between these two views,—evolution without guidance, and evolution with guidance. According to the former, the entire course of things is once and for all determined singly and solely by the initial distribution. It is here that the Laplacean calculator comes in, prepared from the mechanical data of any one moment to find the state of the whole world at any other. For there is one, and only one, course that a system of inert matter will pursue without guidance,—the line of least resistance: it will run down, and it will run down by the easiest and shortest way. But the directions that such a system may be led to take under guidance, but still conformably to the law of conservation, may be innumerable. To forecast the actual progress on this view it is useless to know merely what would happen in accordance with mechanical laws, if the system were left to itself: for any forecast in this case a knowledge of the end or meaning of such progress would be indispensable. Let us take one or two familiar instances by way of illustrating the difference. Imagine a derelict ship and a sea-worthy vessel fully manned: if you know enough of the winds, tides, and currents, you can say where the derelict is likely to be after a week's interval, but this information will be but of secondary importance if you should attempt to predict the position a week later of the ship under sailing orders. Take two trains running opposite ways on a single line of rails, — of which there are hundreds in this country every day: if you know their distance apart, their rates, and that they are left to themselves, you can calculate when and where they will collide. Yet the extreme rarity of collisions is secured simply by what is practically "guidance without work," by 'pointsmen' directing energy which in itself is directionless.

But however impressive the difference between these two forms of process, the blindly mechanical and the intelligently guided, and however surely common sense in our ordinary affairs enables us to distinguish between the two, yet in so far as both are compatible with mechanical principles it is obvious that strictly mechanical considerations will not enable us to distinguish between them. A bullet aimed to hit the mark conforms to the law of projectiles as completely as one fired at random. But now, of a thousand bullets so fired haphazard, probably one or more will hit equally truly. This simple instance may serve to characterise two ways in which the teleological aspect of things can be viewed mechanically. The first is by way of primitive collocations. As the marksman's aim determined the initial movement of the bullet with a view to its final impact on the bull's-eye, so the Creator chose that particular configuration of nebulous matter from which the existing cosmos would mechanically ensue. So Whewell, Chalmers, Jevons, and others represent the beginning of evolution. "Out of infinitely infinite choices which were open to the Creator, that one choice must have been made which has yielded the Universe as it now exists," says Jevons. We may venture, I think, to call this a short-sighted and fatuous view; but I am quite aware that those who first propounded it had many qualifications in reserve, qualifications, however, which must logically resolve the external Artificer into an immanent Spirit. But at all events this half-way house, whatever it be worth, is closed against Mr. Spencer, if even he were disposed to occupy it. For him there can be no 'ultimate properties of kinds,' and no specific collocation of diverse natural agents. Thoroughgoing homogeneity, diffusion, and imperceptibility, are, as we shall see presently, incompatible with such variety in the positions and mechanical endowment of the primitive particles as Jevons, for example, supposes. To Mr. Spencer there is open only the second way of one chance hit out of many misses. We have all of us to admit that facts are by no means wanting that may seem to justify such a view of Nature at least in details, as when finding, for example, "that of fifty seeds, she often brings but one to bear." For the mechanical theory of evolution, however, this second way is absolute and universal. But it will be best here to cite Mr. Spencer's own words: "We have to contemplate the matter of an evolving aggregate as undergoing not progressive integration simply, but as simultaneously undergoing various secondary redistributions; we have also to contemplate the motion of an evolving aggregate, not only as being gradually dissipated, but as passing through many secondary redistributions on the way towards dissipation."[1] Such is Mr. Spencer's general summary; but it would be useless, I fear, to attempt to quote also any of the numerous instances even of physical phenomena, to say nothing of phenomena of a higher order, which he has gathered together in such impressive and bewildering variety in order to substantiate it. I can put the case best, as I understand it, by taking an illustrative instance of my own. Imagine a single drop of water falling alone over Niagara: it will go with accelerated velocity straight from top to bottom. Such a process may typify simple evolution. Now try to realise what happens when the full volume of the St. Lawrence pitches over the Falls. The end is in the main the same as before, but in the course of simple evolution on this larger scale there occur many, and some very striking, instances of compound evolution, in other words, of redistributions, arrests, and reversals of the main process. Individual drops and groups of drops may dash each other into mist, fall, rise, and fall again, eventually joining the stream below only after a long time and by the most devious routes. Imagine the height of the Falls and so the time of falling to be vastly increased, and the secondary results will be more varied still. To this head of compound evolution, then, we are asked to refer all the complex-

[1] *First Principles*, p. 396.

ity of structure and movement, all the varieties of form and rhythm, of which the actual world consists. "Hence," says Mr. Spencer, "other things being equal, in proportion to the quantity of motion which an aggregate contains will be the quantity of secondary change in the arrangement of its parts that accompanies the primary change in their arrangement. Hence, also, other things equal, in proportion to the time during which the internal motion is retained, will be the quantity of this secondary redistribution that accompanies the primary distribution."[1] A little reflection will shew, I think, that on this doctrine what others secure by primitive collocations is secured by taking things on a sufficiently large scale, and trusting to the combinations which haphazard will give. Shuffle an adequate number of fonts of type long enough and a given play of Shakespeare will be among the throws; for it is a possible combination, and in time all possible combinations may be expected. In fact, Mr. Spencer's law of evolution seems to consist essentially in treating the universe as a vast problem in thermodynamics, so to speak.

Apropos of this I cannot do better than quote a striking passage from a letter of Boltzmann's that appeared in *Nature* about a year ago: "We assume that the whole universe is, and rests forever, in thermal equilibrium. The probability that one (only one) part of the universe is in a certain state, is the smaller the farther this state is from thermal equilibrium; but [on the other hand] this probability is greater, the greater is

[1] *Ibid.*, p. 289.

the universe itself. If we assume the universe great enough, we can make the probability of one relatively small part being in any given state (however far from the state of thermal equilibrium) as great as we please. We can also make the probability great that, though the whole universe is in thermal equilibrium, our world is in its present state. It may be said that the world is so far from thermal equilibrium that we cannot imagine the improbability of such a state. But can we imagine, on the other side, how small a part of the whole universe this world is? Assuming the universe great enough, the probability that such a small part of it as our world should be in its present state, is no longer small. If this assumption were correct, our world would return more and more to thermal equilibrium; but, because the whole universe is so great, it might be probable that at some future time some other world might deviate as far from thermal equilibrium as our world does at present."[1]

By 'world' I take Boltzmann to mean what is commonly called the 'visible universe' or 'our galaxy.' The return to thermal equilibrium again corresponds to Mr. Spencer's simple evolution, assuming a like fortuitous initial distribution or absence of specific collocations, and a universe indefinitely great. Of course there is no lack of space and time; even energy too is cheap, when you have only to imagine it. But such a chance oasis of life and order in an illimitable desert of monotonous irregularity is, I need hardly say, *not* what Mr. Spencer means by evolution. So much the worse.

[1] *Nature*, 1894-1895, vol. li, p. 415.

however, for his synthetic philosophy. For while that is the most that his law entitles him to, he assumes not merely that the present throw — to recur to my illustration — is comparable to a play of Shakespeare, but he assumes also that, after the processes of dissolution shall have broken up the type, another play will be thrown next time. In other words, he is guilty of the amazing fallacy of supposing that, because the laws of energy are everywhere present, they are everywhere sufficient to explain what we see; which is much the same as assuming that, because a painter's palette, like his finished canvas, shews us a mixture of colours laid on with a brush, therefore what sufficed to produce the one would equally suffice to produce the other.

But the further exposure of this prime fallacy of Mr. Spencer's synthetic philosophy must be reserved till next lecture.

LECTURE VIII

MR. SPENCER'S INTERPRETATION OF EVOLUTION

Mr. Spencer proposes to deduce the phenomena of evolution (celestial, organic, social, etc.) from the conservation of energy. The obvious insufficiency of this principle taken alone. Mr. Spencer's conception of it contrasted with that of Helmholtz.

How Mr. Spencer connects this 'persistence of force,' as he prefers to call it, with his doctrine of the Absolute. The vagueness of his terms.

The three principles in Mr. Spencer's interpretation: 1. Instability of the homogeneous. *But is the homogeneous necessarily unstable? Quite the contrary. Moreover, Mr. Spencer cannot by analysis get at such a beginning as he supposes. How much can evolution possibly account for, and how little need it presuppose? No clear advance to be made from Mr. Spencer's standpoint. Some illustrative instances of Mr. Spencer's procedure:* (a) *self-rotating nebulæ: in a single homogeneous object no ground of change;* (b) *instability of circular orbits: looseness of Mr. Spencer's terminology;* (c) *chemical differentiation, instability of the* heterogeneous: *two-edged arguments.*

2. Multiplication of effects. *An instance of what Mr. Spencer understands by one cause and many effects. Illusory deduction of this principle from the fundamental one of persistence of force.*

3. Segregation. *This the key to the advance from vague chaotic heterogeneity to orderly heterogeneity.' The process described: it turns out to require only 'forces acting indiscriminately.' Relation of this principle to the other two. Difficulties for Mr. Spencer in connection with the distribution of the chemical elements. Also in the characteristics of organisms and the products of human industry. But Mr. Spencer's terminology is happily 'plastic.'*

As we shall have to refer frequently to Mr. Spencer's formula of evolution in its final form, I will begin by quoting it at length: "*Evolution is an integration of matter and concomitant dissipation of motion* [so much answers to 'simple evolution' and has been quoted already; what follows includes 'compound evolution'] *during which the matter passes from an indefinite, incoherent homogeneity to a definite, coherent heterogeneity; and during which the retained motion undergoes a parallel transformation.*"[1]

"The task before us," says Mr. Spencer, the law of evolution being ascertained, "is that of exhibiting the phenomena of Evolution in synthetic order. Setting out from an established ultimate principle, it has to be shown that the course of transformation among all kinds of existences cannot but be that which we have seen it to be. It has to be shown that the redistribution of matter and motion must everywhere take place in those ways and produce those traits, which celestial bodies, organisms, societies, alike display. And it has to be shown [here is the point] that this universality of process results from the same necessity which determines each simplest movement around us, down to the accelerated fall of a stone or the recurrent beat of a harp-string. In other words, the phenomena of Evolution have to be deduced from the Persistence of Force. As before said, 'to this an ultimate analysis brings us down; and on this a rational synthesis must build up.'"[2]

By Force Mr. Spencer means, among other things, Energy. Now I think it is quite clear that, so far from accounting for all the phenomena of evolution,

[1] *First Principles*, p. 396. [2] *o.c.*, p. 398.

the doctrine of the persistence of energy alone will not account for a single one. The celestial, organic, social, and other phenomena which make up what Mr. Spencer calls cosmic evolution are so many series of qualitative changes. But the conservation of energy is not a law of change, still less a law of qualities. It does not initiate events, and furnishes absolutely no clew to qualitative diversity. It is entirely a quantitative law. When energy is transformed, there is precise equivalence between the new form and the old; but of the circumstances determining transformation and of the possible kinds of transformation the principle tells us nothing. If energy is transferred, then the system doing work loses precisely what some other part of the universe gains; but again the principle tells us nothing of the conditions of such transferences.

As I tried to shew briefly in the sixth of these lectures, this principle may be regarded as primarily and fundamentally logical. Somewhere or other we postulate persistence or conservation, and finding so far as experience goes that mass and energy are conserved, we apply to them this *a priori* postulate. It might turn out that we were wrong in this application, but the postulate in its abstract generality we should still not question. In some sense it must be true to say *Causa æquat effectum*, and meanwhile there is a vast body of evidence to shew that it is true of the transferences and transformations of energy. But now the fact that the principle of energy involves in this wise both an *a priori* and an empirical factor is continually ignored by Mr. Spencer. He lays all the stress on the *a priori* factor, *i.e.* on his own extraordinary version of it; and does not see that this by itself is ludicrously insufficient. Hence such language as this, with which his chapter on the Persistence of Force concludes: "Deeper than demonstration — deeper even than definite cognition — deep as the very nature of mind, is the postulate at which we have arrived. Its authority transcends all other whatever; for not only is it given in the constitution of our own consciousness, but it is impossible to imagine a consciousness so constituted as not to give it."[1] And now let me quote for comparison with this a sentence or two from the conclusion of Helmholtz's famous essay on the same subject: "I believe that what has been here advanced has shewn this law to be contradicted by no facts at present known to science, but to be strikingly confirmed by a very large number. I have striven to exhibit as completely as possible such consequences as follow from it in combination with the laws of natural phenomena so far ascertained, consequences which must still await experimental verification. It has been my aim to lay before physicists with all possible completeness the theoretical, practical, and heuristic importance of this law, the complete establishment of which may well be regarded as one of the chief undertakings of the immediate future."[2] Such language as this would be not only sheer nonsense, but a sheer impossibility if Mr. Spencer's philosophy were right. Clearly Helmholtz does not regard the persistence of force as a

[1] *First Principles*, p. 192.

[2] *Erhaltung der Kraft*, Ostwald's edition, p. 53.

datum of consciousness. But now Mr. Spencer, in a very solemn passage, declares that "if it can be shewn that the persistence of force is not a datum of consciousness," why, "then, indeed," he adds, "it will be shewn that the theory of Evolution has not the high warrant here claimed for it."[1] The burden of proof, however, plainly lies with him. Here is a principle, of which physicists fifty years ago were unaware, a principle which has had to fight its way to recognition, a principle the range of which is still a question — the notion of dynamically non-conservative systems being therefore not absurd; if this principle lies so wondrous deep, "deeper than demonstration, deeper even than definite cognition," then let Mr. Spencer explain Newton's ignorance of it and the general scepticism that greeted its enunciation by Mayer, Joule, and Helmholtz. Perhaps the terrible depth from which they must have brought it is the explanation!

Taking this principle, then, as physicists understand it, and not as it is misunderstood by Mr. Spencer, I repeat that it will not carry us one step towards his evolutionary formula. You could not deduce from it even those "simplest movements," "the accelerated fall of a stone or the recurrent beat of a harp-string," which he assumes to be necessarily determined by it. Yet still more hopeless, if possible, would it be to find for "the theory of evolution the high warrant claimed for it" if we took Mr. Spencer's own version of the persistence of force instead of the accepted doctrine.

To examine this version must appear, I fear, some-

[1] *First Principles*, p. 553.

what of a digression. But let me remind you how often this recognised champion of naturalistic evolution reiterates his confidence that nothing short of a refutation of this ultimate position can shake his general conclusions: "to this," he has said, "an ultimate analysis brings us down and on this a rational synthesis must build up." "But now what is the force of which we predicate persistence?" asks Mr. Spencer; and he answers: "It is not the force we are immediately conscious of in our own muscular efforts . . . the force of which we assert persistence is that Absolute Force, of which we are indefinitely conscious. . . . By the persistence of Force [capital F], we really mean the persistence of some Power [capital P] which transcends our knowledge and conception. The manifestations, as recurring either in ourselves or outside of us, do not persist; but that which persists is the Unknown Cause [capitals again] of these manifestations."[1] In this statement it is important to note two things. First, that between the manifestations or phenomenal forces, according to the usual phrase, and this Absolute Force or Power, there stretches all that gulf, which Mr. Spencer has elsewhere magnified, separating the known and comprehensible from the unknown and incomprehensible. Secondly, that by persistence as applied to phenomenal forces he means the quantitative constancy of these in their totality; while by persistence as applied to Absolute Force he means, as he says, to assert "an Unconditioned Reality, without beginning or end." Now, if Mr. Spencer's cosmic philosophy does not fall between

[1] *Ibid.*, p. 189.

these two supports or lose itself in that 'ugly, broad ditch' Schelling spoke of, between Nature and the Absolute, it will be luckier than most eclectic attempts. If it had started from the Absolute and Unconditioned Reality, of which we are said to be indefinitely conscious, it would obviously have been gratuitous — nay, self-contradictory and nonsensical — to assume that the manifestations of this Unknowable to finite intelligences must remain always quantitatively the same. If "rational synthesis" of things is what we seek, it is surely more reasonable to say with Lotze: "What lies beneath all is not a quantity which is bound eternally to the same limits and compelled through many diverse arrangements, continuously varied, to manifest always the very same total. On the contrary, should the self-realisation of the Idea require it, there is nothing to hinder the working elements of the world being at one period more numerous and yet more intense; at another period less intense as well as fewer. Then would the course of Nature be like a melody, not flowing in monotonous uniformity, but with *crescendos* and *diminuendos* as each in turn is required to express the meaning of the whole."[1]

If now, on the other hand, Mr. Spencer had started from the phenomenal, then, allowing as he does, that of the conservation of energy neither inductive proof nor demonstration is possible, he ought to have regarded that law as, like the still wider law of causation, a postulate or regulative principle connecting together the various branches of physics. But a basis so tenta-

[1] *Metaphysik*, 1879, § 209.

tive and restricted would not suffice for a theory which essays to exhibit all the changes of celestial bodies, organisms, and societies as necessary results of the same universal principle. "The recognition of a persistent Force, ever changing its manifestations but unchanged in quantity throughout all past time and all future time, is," he declares, "that which alone makes possible each concrete interpretation and at last unifies all concrete interpretations."[1] So he is led to perpetrate two or three astounding feats of philosophical jugglery. The apparatus of the first of these we have now before us. Persistence in the sense of *permanence* is secured first of all by reference to the Unconditioned Reality; the non-existence of which is unthinkable, although any knowledge of it is impossible—since to know is to condition. Next persistence, but in the sense of *quantitative constancy*, is transferred from this Unconditioned Reality to its phenomenal manifestations, but only by first affirming of it precisely that statement which we are not empirically warranted to affirm absolutely of them. Let me present this apparatus anew in Mr. Spencer's own words. *Item No.* 1. "Getting rid of all complications and contemplating pure Force, we are irresistibly compelled by the relativity of our thought, to vaguely conceive some unknown force as the correlative of the known force."[2] *Item No.* 2. "Every antecedent mode of the Unknowable must have an invariable connexion, quantitative and qualitative, with that mode of the Unknowable which we call its consequent. For to say otherwise is to deny the per-

[1] *First Principles*, p. 552. [2] *o.c.*, p. 170.

sistence of force."[1] *Item No.* 3. "For persistence is nothing more than continued existence, and existence cannot be thought of as other than continued."[2] In the first we get the absolute existence of Force, with a capital F, at the price of absolute ignorance concerning it; in the second, we get the absolute constancy of force, with a little f, at the price of making precise and definite statements concerning that Unknowable. The intellectual somersault thus rapidly performed is covered by taking continued existence to involve invariable quantity. How quantity of Unconditioned and Unknowable Reality is to be measured we are not told, nor yet what the unit of measure is to be. Does not this step deserve the name of intellectual jugglery: on two items of ignorance to establish an ultimate principle determining what the course of transformation among all kinds of existences must be? We do not *know* what the Absolute is and we cannot *prove* that the quantity of force remains always the same. But since no consciousness can think being as not being, the persistence of the persistent is the fundamental cognition from which all others are derived; hence the Unknowable in persisting must make the knowable that does not persist a constant quantity.

By such fetches of ingenuity to resolve the Absolute into a fixed quantity would, after all, not be worth the pains, unless, as I have said, force is to have a much wider meaning and the conservation of energy a much wider range than science at present allows to them. Otherwise it would be impossible to bring organisms

[1] *First Principles*, p. 193. [2] *o.c.*, p. 195.

and societies and all thereto pertaining — life, mind, character, language, literature, and institutions of every kind — under the cover of a single formula. We are therefore not surprised to find Mr. Spencer treating of the transformation of physical forces into mental forces and insisting on a quantitative equivalence between the two, just as he treats of the transformation of mechanical work into heat and the value in foot-pounds of a calorie. The poetry of Milton and the British Constitution, nay, the human mind and the Christian religion, are all according to him, equally with the tidal bore on the Severn or gales at the equinoxes, so many secondary results of the nebular hypothesis, cases of integration of matter and dissipation of motion in obedience to the persistence of force. It is to encompass all these within one formula that he is tempted to stretch a great physical generalisation beyond all meaning, and to justify his venture by questionable metaphysics concerning Absolute Being. But it will be time enough to deal with the hopeless vagueness of Mr. Spencer's conceptions of "knowable force" as they arise. Meanwhile, having seen how little he succeeds in obtaining for his theory of evolution the high warrant he claims for it, let us turn to some of the details of the theory itself.

At once we make a great descent. We leave behind the Ultimate Cause, Inscrutable Power, Unconditioned Reality, supposed to be indispensable to Mr. Spencer's "rational synthesis." We now find ourselves confronted, as the complete theory requires, by the whole universe in "a diffused imperceptible state." "On setting out," says our guide, "the proposition which comes first in logical order, is that some rearrangement must result: and this proposition may best be dealt with under the more specific shape that the condition of homogeneity is a condition of unstable equilibrium." Or more precisely: "The absolutely homogeneous must lose its equilibrium, and the relatively homogeneous must lapse into the relatively less homogeneous."[1] But this is going too fast. *Il n'y a que le premier pas qui coûte:* so we must be wary here. That homogeneity implies instability is anything but self-evident. For one thing, if such were the case, it would be difficult to see how, on Mr. Spencer's theory, such homogeneity could ever arise. Any given era of evolution we are free to regard, according to his principles, as preceded by an era of dissolution, the persistence of force being supreme throughout. We seem required to picture the whole universe, as soon as evolution is complete, beginning to decompose and continuing so to do in such a manner that the state of homogeneity shall be simultaneously reached by every part of it. Otherwise, owing to the instability of the homogeneous, the counter-process of redintegration would begin in one part before the others were ready. There seems, however, but one way in which such a simultaneous dissolution is possible, viz.: by the precise and instantaneous reversal of every movement throughout the whole, as stated, *e.g.*, in the passage from Helmholtz quoted in the last lecture. The universe would then be like a reversible musical box which could play its tunes backwards; and, assuming it to have started from a homogeneous state, it would in

[1] *First Principles*, pp. 400, 429.

this way return to it. But this is not what Mr. Spencer understands by dissolution. In truth, however, homogeneity is not necessarily instability. Quite otherwise. If the homogeneity be absolute, — that of Lord Kelvin's primordial medium, say, — then the stability will be absolute too. In other words, if "the indefinite, incoherent homogeneity," in which, according to Mr. Spencer, some rearrangement *must result*, be a state devoid of all qualitative diversity and without assignable bounds, then, as we saw in discussing mechanical ideals, any "rearrangement" can result only from external interference; it cannot begin from within. All physicists are agreed, as Messrs. Tait and Stewart put it, that "in the production of the atom from a perfect fluid, we are driven at once to the unconditioned — to the Great First Cause; it is, in fine, an act of creation and not of development."[1] Thus, the very first step in Mr. Spencer's evolution seems to necessitate a breach of continuity. This fatal defect is not apparent in his exposition; but only because, as remarked in the last lecture, the whole vast problem of molecular development is lost in the haziness of the nebular theory; and, further, as we now see, is slurred over by the vagueness of such terms as "indefinite, incoherent homogeneity." Mr. Spencer's attempt to evolve the chemical elements from prime atoms by means of the nebular hypothesis has, I believe, impressed nobody — unless it be with his failure to realise the endless complications with which such a problem is beset. But suppose this stage of evolution satisfactorily explained, still what of the prime

[1] *The Unseen Universe*, second edition, p. 117.

atom? Are we to call that indefinite, incoherent, homogeneous? How can an atom be indefinite or incoherent? How, then, if we are to begin with the indefinite and incoherent, can we begin with an atom of any sort? And if we go beyond atoms to some cosmic protyle such as that of Sir William Crookes, must we not assume, too, as he suggests, that this "elementary *protyle* contains within itself the potentiality of every possible combining proportion or atomic weight,"[1] and then how can it be homogeneous? There is, however, no end to such questions. At any rate our reflections on the kinetic ideal of matter brought us, it may be remembered, to this conclusion.

That conclusion suggests two or three further remarks on Mr. Spencer's "interpretation of evolution." In the first place, the synthetic philosophy cannot begin at the beginning of evolution because physical analysis can never place it there. Such conceptions as prime atoms, primordial media, *prima materia*, and the like, are obviously ideal limits and not possibly presentable realities. In the next place, such limiting conceptions, taken alone and treated as realities, lead straightway to absurdities. We cannot begin operating with zeros and infinities, though we recognise quantities that approximate to them asymptotically. So, in like manner, qualitative diversity may be replaced by quantitative formulæ and the range of mathematical description extended without assignable limit. But such procedure is plainly one of abstraction, and — if carried to the uttermost — leaves us, as we saw, with absolutely no real content to

[1] Address at Brit. Assn., 1886, *Nature*, vol. xxxiv, p. 428.

which our numbers and diagrams apply. A real world is sublimated into "non-matter in motion." To such epistemological reflections our synthetic philosopher seems altogether a stranger, or he could never have perpetrated the transparent absurdity, doubly an absurdity in his case, of representing any heterogeneity as arising simply—provided only there is quantitative equivalence—out of absolute homogeneity. Such homogeneity is essentially stable; and thus the first step in his scheme of evolution becomes impossible, because, in his zeal to be thorough, our author has eliminated all ground of difference. Or if he has not, he has failed to make good his undertaking, and begins not at the beginning, but with atoms having indefinitely many potentialities and distributed according to some specific configuration; in other words, begins with the manufactured articles of Herschel and Maxwell, and the collocations of Chalmers and Mill. In point of fact he begins, as said, with the 'nebular hypothesis,' all that goes before it being adroitly covered by the utterly unscientific and unphilosophical phrase 'indefinite incoherent homogeneity.'

One further remark before we proceed; the proposal to start with complete homogeneity leads us to ask: How much can evolution possibly account for, and how little need it presuppose? According to Mr. Spencer's drift, it would seem that evolution, expounded in thorough, philosophical fashion, will account for all form, provided only a fixed quantity of matter and energy is given. As Professor Riehl humorously puts it: "Listen to Herbert Spencer and you must believe that literally everything there is has evolved, including forsooth even evolution itself."[1] But so long as we look at things from a purely mechanical standpoint, as Mr. Spencer does, it is difficult to see what ground there is for asserting any increase of complexity at all. Given a certain aggregate of mass-points regarded as a conservative system, and there will be a certain number of possible configurations through which it can pass; but on what grounds, I would ask, is one to be called more homogeneous or more heterogeneous than another? "The portions of which the whole is made up may be severally regarded as minor wholes,"[2] says Mr. Spencer. No doubt they may be, but all such individualisation is, from a strictly mechanical standpoint, purely arbitrary. There may be teleological reasons in plenty, or what we may call methodological reasons, or reasons of practical interest; but all such grounds as these transcend the level of Mr. Spencer's primordial truth and its corollaries. Keeping strictly to that, there is only one true homogeneity, the homogeneity of an undifferentiated plenum such as Descartes or Lord Kelvin supposes. Between such a plenum and an aggregate of elements in motion there is no continuity; to secure the differentiation that an aggregate implies, a catastrophe is indispensable. But once such a system is secured, it is meaningless to call it indefinite or incoherent. A configuration cannot be indefinite; and as the forces between every pair of elements depend solely on their masses and positions, such a system

[1] *Der philosophische Kriticismus*, Bd. ii, 2te Th. p. 75.

[2] *First Principles*, p. 428.

is never incoherent, that is to say, is never disconnected. To the Laplacean calculator, *i.e.* according to what Thomson and Tait call 'accurate mathematical investigation' by 'the only perfect method,' a chunk of granite or even a whirl of dust may be just as definite, just as connected, just as heterogeneous as a chronometer or a balance, just as much a pure mechanism conforming to the laws of energy. Summing up on this head, then, we may say: (1) That this opposition of homogeneity and heterogeneity is essentially out of place in a rigorously mechanical theory. (2) That on such a theory it is impossible to interpret Mr. Spencer strictly when he says, "The absolutely homogeneous *must* lose its equilibrium and the relatively homogeneous *must* lapse into the relatively less homogeneous";[1] for instability is incompatible with absolute, and independent of relative, homogeneity. (3) That mere indefiniteness and incoherence entitle him to assert nothing either concerning homogeneity, or stability, or anything else.

Any one at the trouble to read at all critically the long chapter devoted to this so-called Instability of the Homogeneous, cannot fail to discover instances in plenty of what I say. Mr. Spencer's main example I may perhaps be allowed to mention, though it has been already exposed;[2] for in this he flatly contradicts the very mechanical principles he has declared to be so

[1] Spite of this Mr. Spencer, in an earlier foot-note, cuts away the ground from under his own feet by bargaining that "the terms here used must be understood in a relative sense." Cf. p. 330; as also p. 407 *fin.*

[2] Cf. *British Quarterly Review*, 1873, vol. 58.

unfathomably fundamental. Having by a series of gratuitous and sometimes erroneous suppositions got from pristine homogeneity as far as "irregular masses of slightly aggregated nebular matter" all in motion, he continues thus: "Established mechanical principles . . . justify the conclusions that the motions of these irregular masses . . . towards their common centre of gravity must be severally rendered curvilinear, by the resistance of the medium from which they were precipitated; and that in consequence of the irregularities of distribution already set up, such conflicting curvilinear motions must, by composition of forces, end in a rotation of the incipient sidereal system." Now this is a gigantic and palpable blunder, one that even the least mathematically-minded might have avoided by reflecting that matter being essentially inert can hardly be conceived to set itself spinning merely because there is plenty of it. This felicitous plan for securing the rotation that Laplace was content to assume, is directly counter to what is technically called the Conservation of Angular Momentum; and this, it is well known, is directly deducible from Newton's third law. Now the odd thing is that Mr. Spencer — very inaccurately, to be sure — identifies the said law, viz., that action and reaction are equal and opposite, with the law of the conservation of energy. Thus an important scene in his evolutionary drama runs directly counter to its main motive.

Of course Mr. Spencer has had no difficulty in finding instances in plenty of comparatively homogeneous states lapsing into more heterogeneous ones; and had he so minded he could have found just as many instances of 'heterogeneous' states lapsing into more 'homogeneous' ones — as he does indeed when he wishes to illustrate dissolution. And all such instances alike are conformable to the principle of the conservation of energy; thereby shewing, as we have already seen, that that principle is a sufficient basis for none. Whether an egg is transformed into a chicken, into an omelette, or into rottenness, one change is as much, or as little, as the other deducible from that persistence of force which Mr. Spencer always mentions with such mystic awe. Moreover, all such instances require that besides the homogeneous and unstable object or the heterogeneous and unstable object, as the case may be, there should be external forces affecting it. An egg alone in the void would neither hatch nor cook nor smell: it is on the object + external causes that the result — be it more, be it less complexity — essentially depends. Now the universe, regarded as a single object and homogeneous, has no environment, is not amenable to extraneous forces — a peculiarity that makes Mr. Spencer's instances rather refute than corroborate his main thesis, but confirms on the other hand the antithesis we have opposed to it.

Perhaps the most striking thing about Mr. Spencer's multitudinous illustrations of the transitoriness of all things homogeneous and their inevitable lapses into heterogeneity, is the looseness with which these terms are used. Thus he chooses to regard a circular orbit as homogeneous and elliptic orbits as heterogeneous, and then remarks: "All orbits, whether of planets or satellites, are more or less excentric . . . and were they perfect circles they would soon become ellipses. Mutual perturbations would inevitably generate excentricities. That is to say, the homogeneous relations would lapse into heterogeneous ones."[1] Now in the first place let an orbit be what it may, the relations determining it are invariable, involve no more factors at one time than at another. But even if an orbit could with propriety be called a relation, it is especially absurd in Mr. Spencer to contrast a circle which is a single figure with ellipses of which there may be an indefinite multitude. Compare an elliptic orbit of definite eccentricity with a circular orbit, which is itself an elliptic orbit of definite, i.e. *zero*, eccentricity, and both appear equally homogeneous and equally stable. Apropos of this a mathematical critic of Mr. Spencer, after comparing him to a man "who thought that Nature had a spite against the figure 3, because he had noticed that it was much more usual to find that a number did not end with 3 than that it did," proceeds to remark: "Of course, if you put all elliptical orbits in one class and leave the circle to form another class by itself, it is likely that the orbit will tend to belong to the first-named class; for it can change through all possible ellipses without altering the *appellation* of its orbit, while the slightest variation from a circle is reflected in a change of name."[2] A blunder of this kind, though it shews how flimsy Mr. Spencer's constructions are, would scarcely be worth mention if it were isolated. Unhappily the fallacy

[1] *First Principles*, p. 410.

[2] *British Quarterly Review*, article above referred to.

underlying it is general and vitiates an indefinite number of 'the great evolutionist's' arguments; for the homogeneous is ever one and the heterogeneous always many.

Yet another instance may be mentioned in view of its subsequent importance. Mr. Spencer devotes one section of his long chapter on the Instability of the Homogeneous to what he calls "chemical differentiations." In the course of it he illustrates the well-known, but for his argument somewhat anomalous, fact that in general "simple combinations can exist at a higher temperature than complex ones," in other words that chemical stability decreases as chemical complexity increases, so that for example what we ordinarily regard as chemical elements at one extreme cannot be decomposed by any heat that we can artificially produce, whereas organic compounds at the other extreme, which are extremely complex, are readily decomposed at quite moderate temperatures. Now as all ponderable matter is in some chemical state or other, and as the half of our evolutionary formula relates to redistribution of matter, this fact — that chemically the more homogeneous matter is the more stable — surely cuts a monstrous cantle out of the best of Mr. Spencer's realm.[1] I say the best, for here, at any rate, the terms homogeneous and heterogeneous are strictly applicable. The strange thing, however, is that when, in a subsequent volume of

[1] Our author elsewhere (§ 101) accounts for this greater stability of what is chemically homogeneous by asserting the comparative absence from it of "contained motion." But even this surrenders the point that the homogeneous, merely as homogeneous, is unstable. It suggests, indeed, the precisely opposite conclusion.

his philosophy, Mr. Spencer comes to treat of the evolution of organic life, this instability of the ***heterogeneous*** becomes the mainstay of his argument.

But why, you may wonder, does he bring it forward in a general chapter that has to prove the instability of the homogeneous, where it seems so irrelevant and inopportune? It is the earth's crust which is here the direct object of Mr. Spencer's exposition: his purpose, he says, is "to show how, in place of that comparative homogeneity of the earth's crust, chemically considered, which must have existed when its temperature was high, there has arisen during its cooling, an increasing chemical heterogeneity, each element or compound, being unable to maintain its homogeneity in presence of various surrounding affinities, having fallen into heterogeneous combinations."[1] Let us examine this argument for a moment. If the comparatively homogeneous as such is unstable, then *a fortiori* the altogether homogeneous should be unstable, if the argument is to be worth anything. Let us then, as we surely may, imagine the incandescent globe to have been wholly of oxygen or of silicon, ought we not then to expect that heterogeneous combinations would appear sooner and more conspicuously? Again if the instability is due to homogeneity simply, why is it essential to reduce the temperature and to insure "the presence of various surrounding affinities" before the lapse into heterogeneity can begin? Further, if the homogeneity involves instability, how comes it that once combination has begun "the stability *de*creases as the complexity *in*creases?" Lastly, what

[1] *First Principles*, p. 411.

warrant has Mr. Spencer for saying that 'each element or compound falls into combination, being unable to maintain *its* homogeneity'? Does he mean that, when oxygen and hydrogen form water, or acid and base form a salt, both components disappear? How then can the combination be called heterogeneous; we should surely have a new homogeneous, presumably as unstable as before! On the whole I think we may say that while Mr. Spencer's main argument here is an instance of the "indefinite incoherent" confounding of things in themselves distinct, it incidentally lets in a whole flood-gate of facts very damaging to the homogeneity of his theory.

With other instances of the instability of the homogeneous supposed to be deducible from the persistence of energy, such as the development of intelligence and the desynonymisation of words, it is impossible to deal here. Mr. Spencer is considerate enough to anticipate his readers' misgivings so far as to assure them that "any difficulty felt in understanding" these and like instances "will disappear on contemplating acts of mind as nervous functions." All such parts of Mr. Spencer's doctrine, then, may for the present stand over. There remain still two steps in what our cosmic philosopher calls the *rationale* of evolution, its deduction, that is to say, from the persistence of force. At each of these we must glance briefly.

To secure his first step, Mr. Spencer, as we have seen, was led to maintain that the homogeneous is essentially unstable; his second step consists in maintaining that "the effect is universally more complex than the cause."[1] "This secondary cause of change from homogeneity to heterogeneity," he remarks, "obviously becomes more potent in proportion as the heterogeneity increases,"—in fact, "the multiplication of effects," as he entitles his second step, must, he contends, "proceed in geometrical progression. Each stage of evolution must initiate a higher stage." All these conclusions, of course, he proceeds as before to shew, "are not only to be established inductively, but are to be deduced from the deepest of all truths."[2] And again I can only contend that strictly interpreted this second position is as devoid of foundation as the first, and is only made to look plausible by a very loose use of leading terms and a superabundance of specious analogies.

Let us see, for instance, what Mr. Spencer means by one cause and by many effects. Here is an example. He gives a detailed description of the leading physical features of the earth,—its mountain ranges, irregular coast line, its continents, and its oceans; and then concludes by saying: "Thus endless is the accumulation of geological and geographical results slowly brought about by this one cause—the escape of the earth's primitive heat."[3] The effects, no doubt, are multitudinous enough, but on what ground is the cause accounted one? Suppose the earth to be a single gas cooling under constant pressure, or to consist entirely of one pure metal—the escape of the primitive heat could take place as before, but how many of the endless effects of this one cause would there be left? If such loose and popular language is to pass as scientific induc-

[1] *First Principles*, p. 433. [2] *o.c.*, p. 458. [3] *o.c.*, p. 438.

tion, it would be every whit as easy to shew that a single effect is due to a multiplicity of causes. The historian, for example, may in all seriousness so regard the Reformation or the French Revolution, and the more patient and pertinacious he is the more multitudinous the causes he will find for that one result.

But when causes and effects are to be deduced from a quantitative law and expressed in terms of matter and motion, we have a right to expect more precision. Mr. Spencer begins by using the language of the exact sciences, talks much of incident forces, of action and reaction being equal and opposite, and so forth, but in the end he is as careless as one quite ignorant of mechanical principles. Thus, for instance, when he describes the fracture of a stone by a hammer as a case in which a single force is changed by conflict with matter partly into forces differing in their directions and partly into forces differing in their kinds. He then proceeds further to describe the first of these as a change of a homogeneous momentum into a group of momenta, heterogeneous in both amounts and directions. Lastly he mentions as instances of the second the sound produced, the heat disengaged, and the sparks struck off, etc. In the course of half a page force is used in three different senses — as mechanical energy, as momentum, as a physical sense-impression — and all wrong. But, above all, what is to be understood by "a conflict of force with matter?" To the physicist proper, Professor Tait say, for whom matter is essentially passive and inert, such language is nonsense; it can hardly have more meaning for a writer who, like Mr. Spencer, maintains that matter is force and nothing else. How, we wonder, by the way, did the homogeneous lapse into this kind of heterogeneity?

From the inductions, of which these are specimens, Mr. Spencer next passes to the deduction of this second step from "the deepest of all truths," and in so doing he becomes suddenly very perfunctory. After the parallel deduction in the case of his first step, a like argument, he thinks, "seems here scarcely required," and he is content "for symmetry's sake briefly to point out how the multiplication of effects, like the instability of the homogeneous, is a corollary from the persistence of force." In less than two pages the thing is done, or rather not done, not even attempted — a result which in view of the flimsiness of the inductive argument is only to be regretted. What Mr. Spencer has to prove can be stated simply enough. It is that if "the quantity of Force remains always the same," there must be, and unless the quantity of Force remains always the same, there cannot be, what he calls the multiplication of effects in geometrical progression. What he actually does, however, is merely to draw out with needless parade a proposition, which, as he is frank enough to allow "is in essence a truism," viz., that unlike causes, or, as he prefers to say, 'unlike forces' will have unlike effects. To this he merely appends the remark that each different modification "must produce its equivalent reaction; and must so affect the total reaction. To say otherwise is to say this differential force will produce no effect, which is to say that force is not persistent."[1] In a word, instead

[1] *First Principles*, p. 457.

of shewing that, given the persistence of energy, there must be this geometrical increase in the diversity of effects, what Mr. Spencer does is to assert that given this diversity, every effect is the equivalent transformation of its cause — which is not to deduce anything as a consequence of the law of conservation; it is only a needless reiteration of the law itself.

We come at length to the final step in the *rationale* of evolution. Mr. Spencer devotes to it the last chapter of his exposition of this subject, and his opening sentences ought to surprise us: "The general interpretation of Evolution," he begins, "is far from being completed in the preceding chapters. . . . Thus far no reason has been assigned why there should not ordinarily arise a vague chaotic heterogeneity in place of that orderly heterogeneity displayed in Evolution." "We have found . . . that the homogeneous must lapse into the heterogeneous and that the heterogeneous must become more heterogeneous." "But," says our oracle, "the laws already set forth furnish no key to this arrangement in so far as it is an advance from the indefinite to the definite." As to the advance from the incoherent to the coherent the key to this, we must suppose, is furnished by that 'simplest and most general aspect' of evolution to which Mr. Spencer ascribes the mere integration or aggregation of matter. But there is, it seems, a further "local integration" or segregation of like from unlike in the heterogeneous mixture. Now it is by this process that orderly heterogeneity arises out of the vague and chaotic. Surprised as we naturally are to find ourselves thus near to the close of the great interpretation, and yet not out of the range of chaos, we await with some anxiety the rationale of this final step by which at the last moment a cosmos is secured. "The rationale," says Mr. Spencer, "will be conveniently introduced by a few instances in which we may watch this segregative process taking place."[1]

Let us be content with one and that the briefest of these instances: "In every river we see how the mixed materials, carried down, are separately deposited—how in rapids the bottom gives rest to nothing but boulders and pebbles; how when the current is not so strong, sand is let fall; and how, in still places, there is a sediment of mud."[2] After this and other introductory instances and an assurance that there are countless similar ones, we have the following generalisation: "In each case we see in action a force which may be regarded as simple or uniform—fluid motion in a certain direction at a certain velocity. . . . In each case we have an aggregate made up of unlike units—unlike in their specific gravities, shapes, or other attributes. . . . And in each case these unlike units or groups of units, of which the aggregate consists, are, under the influence of some resultant force, *acting indiscriminately on them all*,[3] separated from each other—segregated into minor aggregates, each consisting of units that are severally like each other and unlike those of the other minor aggregates."[4] Thus we see that even the transition from the indefinite to the definite, from the vague and chaotic to the cosmical and orderly, is assigned to "force acting indiscriminately."

[1] *First Principles*, p. 459. [2] *o.c.*, p. 460. [3] Italics mine.
[4] *o.c.*, p. 461.

Mens agitat molem is a maxim for which the mechanical theory of evolution has nowhere a place. It is at any rate satisfactory to come to the end and be clear on this point. But I must defer general reflections till the next lecture. For the present let us be content with briefly considering how this indiscriminate sifting process will work in conjunction with the other two.

A good deal will depend on their respective intensities, how they are matched against each other; for it is obvious that in several respects the process of segregation will counterwork the two other causes of evolution. Thus, in producing local integrations of like units, it must act counter to the instability of the homogeneous, according to which the like lapses into the unlike. In so doing, again, it will frustrate the multiplication of effects within the limits of such local integration, for this is efficient 'in proportion as the parts are unlike.'[1] Imagine segregation to have been in full play while the existing chemical elements of the solar system, though present in the nebula, were still uncombined, and that in consequence these elements were separated into minor aggregates severally like each other and unlike the rest — those of high specific gravity or strong physical likeness near together and the unlike far apart. Thus the conceit of the alchemists that the seven metals correspond to the seven planets might have been realised; and as to the gases, oxygen, hydrogen, nitrogen, indispensable constituents of living things — they might have been sifted off into space before planetary consolidation began. We know of

[1] *First Principles*, p. 458.

course that this has not happened or we should not be here. But if Mr. Spencer's principle of segregation is really the potent factor in evolution that he takes it to be, it is at least remarkable to find that with a whole nebula as a field for its activity and untold ages in which to work, it has nevertheless left no trace of itself. Let me quote an excellent authority. "We do not find them [*i.e.* the chemical elements]," said Sir William Crookes in his British Association Address, "evenly distributed throughout the globe. Nor are they associated in accordance with their specific gravities, the lighter elements placed on or near the surface and the heavier ones following serially deeper. Neither can we trace any distinct relation between local climate and mineral distribution. And by no means can we say that elements are always or chiefly associated in nature in the order of their so-called chemical affinities: those which have a strong tendency to form with each other definite chemical combinations being found together, while those which have little or no such tendency exist apart." Then definitely raising the question,—but without any reference to Mr. Spencer, let me say,—"Is there any power which regularly and systematically sorts out the different kinds of matter from promiscuous heaps, conveying like to like and separating unlike from unlike?" this distinguished chemist answers: "I must confess that I fail to trace any such distributive agency, nor indeed, do I feel able to form any distinct conception of its nature."[1] Perhaps Mr. Spencer might have had something to say to this, but

[1] *Nature*, 1886, vol. xxxiv, p. 425.

then, as I have already mentioned, he had stereotyped his *First Principles* some twenty years previously.

One other point as to the relation of the two chaotic or differentiating processes to this cosmic, selectively integrating, principle. At first blush the situation reminds us of that intellectual guidance referred to in the last lecture, when we were distinguishing teleological from mechanical evolution. When human ingenuity constructs a machine or a house, or when Maxwell's sorting demon separates molecules moving with more than average velocity from those moving with less, the processes are what Mr. Spencer might call processes of segregation and local integration. But they differ from Mr. Spencer's process in several respects. First, the result is secured, not by a force acting indiscriminately, but by intelligence counterworking the downhill trend of energy towards dissipation. Also in the case of the products of human skill the result is rather that unlike things are brought together than that unlike things are separated. Nowhere do we find so little segregation, in the sense of Mr. Spencer's sifting and winnowing processes, as in living organisms and the products of human industry. Lastly, organisms and machines are not aggregates of aggregates, but individuals consisting of members. Spite of these essential differences, Mr. Spencer, no doubt, thinks mechanical segregation will cover both, and it must be confessed that by sufficient license in the use of the term 'force' and the free substitution of unit for fragment, individual for aggregate, and the like, the task is feasible, — and the result quite worthless. When sparks rise and dust falls we say each moves along the line of least resistance, their densities and gravitation being the segregating forces; and when the virtuous man 'rises' and the vicious 'falls' we may, if we like, say again that each follows the line of least resistance, and may call their desires and public opinion the segregating forces. This is what Mr. Spencer does like to do; it is what he calls synthetic philosophy.

LECTURE IX

REFLECTIONS ON MR. SPENCER'S THEORY: HIS TREATMENT OF LIFE AND MIND

The conclusions to which we were led in examining the mechanical theory apply here. It is impossible to get more out of a theory than there is in it. Out of space, time and mass, however manipulated, progress, development, history, meaning, can never be deduced.

How has Mr. Spencer come to think this possible? His procedure illustrated. He succeeds by means of formularies that seem to have always a strictly *mechanical sense, though they are frequently only* figuratively *mechanical. Indeed, he outvies the mechanical theorists by his more fundamental analysis as well as by his completer synthesis. But he confounds abstraction with analysis; and abstracts till he has no content left. The eliminated elements are then gradually resumed under cover of the principle of continuity. The existing gaps in scientific knowledge help to cloak such recoveries.*

An instance in Mr. Spencer's transition from Inorganic Evolution to Organic Evolution. Two volumes of the Synthetic Philosophy *missing.*

Mr. Spencer's somersault in passing from Life to Mind. After all, the interpretation of Spirit in terms of Matter is allowed to be 'wholly impossible.'

I HAVE called Mr. Spencer an eclectic. His synthetic philosophy is made up of Hamilton's theory of the Unconditioned, of the physical theory of the conservation of energy as expounded by Grove, of the nebular hypothesis of Laplace, and of what used to be called the development hypothesis, or the doctrine of the transmutation of species. The Darwinian form of this doctrine came too late to be satisfactorily incorporated in his system, still Mr. Spencer was not slow to turn it to account as far as he could. Of his work Darwin, writing to one of its chief exponents, Professor Fiske, thus expresses himself: "Such parts of H. Spencer as I have read with care impress my mind with the idea of his inexhaustible wealth of suggestion, but never convince me; and so I find it with some others. I believe the cause to be the frequency with which I have found first-formed theories erroneous."[1] In passing presently to this narrower subject of biological evolution, I do not propose to refer so fully to Mr. Spencer's views.

In the existing state of knowledge this topic of biological evolution is widely different in subject-matter and methods from the cosmological speculations into which Mr. Spencer attempts to frame it. Here we deal with concrete objects and a vast collection of empirical observations concerning them. The axioms of physics and its ideal conceptions of atoms, ethers, and the like have to be left aside, temporarily at all events. We are forced back upon them again when the dominant naturalistic explanation of the relations of life and mind to their so-called "physical basis" confront us. But having reached a dividing line of this magnitude, it seems appropriate, before proceeding, to attempt a retrospective summary of Mr. Spencer's cosmological presentment of evolution as a deduction from mechanical principles.

It was open to us perhaps to urge at the very outset that such an enterprise was foredoomed to failure. For

[1] *Life and Letters of Charles Darwin*, vol. iii, p. 194.

what Mr. Spencer essays to do is to set before us "the entire history" of things, "considered individually or in their totality"; and to set all this before us as the direct and necessary consequence of a principle of permanence which gives no clue to processes, transformations, or changes of any kind — to say nothing of furnishing the rationale of all processes and changes of every kind whatever. It is as if we had the philosophy of Heraclitus deduced from the premisses of Parmenides. Even when we allow Mr. Spencer to substitute the entire body of hypotheses constituting abstract dynamics for his Eleatic principle of "the impossibility of establishing in thought a relation between Being and Not-being,"[1] the case is not mended. True this transcendent but rather empty principle is not equivalent to the physical doctrine of the conservation of energy; and again the conservation of energy, so far from constituting the sole and sufficient foundation of physical science, only furnishes one of several equations — to put it technically — by which a given transformation is determined. But even if we add to it the principle of least action and all the hypotheses necessary to make a mechanical 'interpretation' of things as complete as such an interpretation can be, still it will be hopelessly inadequate to the "entire history of things considered individually and in their totality." In fact, the conclusions to which we were led in examining the mechanical theory must apply straightway to what is itself but an application of that theory — the resolution of all history into "a total and all-pervading process of redistri-

[1] *First Principles*, p. 191.

bution of matter and motion." It is impossible to get more out of such a theory than there is in it. Between one stage of the process and another there can only be such differences as dynamical diagrams, time-charts, hodographs, and the like will give. The entire history of things would thus be nothing better than the monotonous uniformity of a long series of gigantic Nautical Almanacs. Change there would be certainly, but only change of motion, change of grouping of unchangeable elements, unchangeable because utterly devoid of qualitative diversity or internal character. Progress, development, history, meaning — of these there would be nothing. It is obviously impossible to get such conceptions out of space, time, and mass, as quantities; or out of any relations between them, for these in turn are only quantities. We have only the night — to appropriate a *mot* of Hegel's — when all cows are black. Everything is dynamical diagram: to this common level "celestial bodies, organisms, societies" will all alike have somehow to be reduced.

But how then does Mr. Spencer deceive himself into imagining that he finds increasing purpose, advancing harmony, final perfection, what he is pleased to call a "Philosophieo-Religious doctrine,"[1] in a purely quantitative scheme; a scheme to which all such notions as adaptation, perfection, and happiness are absolutely disparate? The answer is simple and the fallacy to which it has led is clear. There are two points to notice. First, like the rest of us, Mr. Spencer sets out from the concrete world which is only intelligible to us so

[1] *First Principles*, p. 557.

far as we can regard it as a world of individuals, a world full of purpose and of adaptations, a world to which such notions as worth, progress, and perfection are applicable [1] Looking at this world, then, historically, we can range its facts in an ascending order of complexity and value — physical, biological, psychological, social, and so forth. But as we make this ascent we have at every advance to take up new conceptions: the facts of biology cannot be expressed in purely physical terms; psychology will not resolve into biology nor sociology into psychology. It would be sheer waste of time to enlarge upon a point so perfectly obvious, though for any attempt at a theory of knowledge it is a point of vital importance. For Hegel — who also was an evolutionist, but one occupying a standpoint the diametrical opposite of Mr. Spencer's — the exhibition of this hierarchy of categories was the main problem; for Mr. Spencer it is no problem at all. His works testify on every page that such an ascending scale of conceptions is there and unavoidable. But the fact gives him not a moment's pause; it is only one more instance of the passage of matter from indefinite, incoherent homogeneity to definite coherent heterogeneity!

And so we come to the second point, and this again it is enough barely to mention. Whatever be its meaning, its purpose, or its life, the cosmos in one aspect is but matter in motion. However difficult to formulate without appearing to prejudge the ancient and

[1] "Constituted as the human mind is, if nature be *not* interpretable through these conceptions, it is not interpretable at all." Sir J. Herschel on *The Origin of Force* in the *Fortnightly Review*, vol. i, p. 442.

obstinate problems to which it has given rise, this fact is none the less in itself both familiar and unquestionable. The world of ideas is in some way presented through, and embodied in, the world of sense; and the sensible can be summarised in terms of matter, motion, and force. And now it is by his mode of dealing with these two planes of thought that Mr. Spencer has deceived himself into thinking that he has encompassed the entire history of things within the scope of a materialistic formula. He *advances* by way of the ascending scale of ideas, the concrete progress from physics to life, from life to mind, from mind to reason; but he professes to *explain* by falling back on the abstractions of pure dynamics. Yet on this level, if we could imagine ourselves confined to it, there is, as I have frequently urged, no real advance, no true evolution at all. Space and time, of course, do not alter; also mass-elements do not alter; and so between one configuration, one diagram, and another, of a given number of such elements, there is no essential difference. But when we command *both* the dead level of changing configurations and *also* the ascending complexity of the concrete sciences and their categories, then we may make a shift to call one material system a pumpkin and another a poet. Only however because we first know pumpkins and poets as such. To suppose then, that the transformation of one such configuration into another furnishes any clue to the evolution of poets is a glaring and ridiculous blunder. But it is for this blunder that Mr. Spencer is vaunted by Tyndall as an "Apostle of the Understanding whose ganglia are sometimes the seat of a nascent poetic thrill."[1] Let me try to make this point clearer by means of an imaginary case involving the same sort of fallacy. Take a shelf of miscellaneous books in the English language, — books on mathematics, chemistry, physiology, history, art, literature, or what you will, — and imagine a private student setting to work to improve his mind, as we say, by means of them. It will not be indifferent in what order he reads: to understand the physiology he will often find himself thrown back on the chemistry, to understand the chemistry he must often consult the mathematics; the art and the literature will be full of allusions to the history. Above all, the whole will presuppose that the student himself is a person with sense, intelligence, feeling, conscience. Nevertheless, if we are not to be too severe on the synthetic philosophy, we had better leave the student, as much as may be, out of account.

Now let us introduce a man of letters with a Spencerian sense of thoroughness. The first step, he will say, must be to analyse all this material; and only an ultimate analysis will suffice: we must not pause till we have reached the imperceptible. Specialists, he will continue, have already provided nomenclatures and terminologies, glossaries, indexes of persons and things, and the like. Passing beyond all this un-unified knowledge, the lexicographer provides us with partially unified knowledge, and covers the whole range of these books by an adequate dictionary of the English tongue. We get still nearer to that ultimate analysis that must

[1] Belfast Address before the British Association, 1874, p. 49.

underlie completely unified knowledge when we can exhibit the letters of the alphabet as the constituents of English as it is, was, and will be. But even these letters are made up of strokes of two kinds, viz., straight strokes and curved strokes; and only when these are disintegrated into the primordial dots of which they must be compounded, and these dots duly dissipated, have we reached that ultimate and imperceptible state where rational synthesis must begin. Evolution then arises as this dissipation gives place to concentration, and with increased concentration comes increased differentiation; and so we advance from dots to strokes, from strokes to letters of various forms, from these to syllables "with a subsequent advance to dissyllables and polysyllables and to involved combinations of words"—the heterogeneity steadily increasing in geometrical progression. As these aggregates of letters grow in complexity and definiteness more wide-reaching interdependences become manifest: in short, what is called grammar and sense arise. But not only do we find in these the same processes of integration, differentiation, and segregation already exemplified; they are also themselves objectively presented and more or less permanently registered in literal form. Then, when at length the change which evolution presents is complete and equilibration is reached, we have, in what we know as stereotype, that perfection, harmony, and complete congruity which the stereotyped editions of the synthetic philosophy so admirably illustrate. To be sure, this interpretation of all literary phenomena in terms of integrated black and diffused white is nothing more than the reduction of complex phenomena to their simplest forms, and as that philosophy shews "when the equation has been brought to its lowest terms the symbols remain symbols still."[1] No doubt, "most persons," as the author of that philosophy remarks, "have acquired repugnance to such modes of interpretation." But, as he further truly says, "whoever remembers that the forms of existence [in his case Matter and Motion, in ours print and paper] which the illiterate speak of with so much scorn are shewn to be the more marvellous in their attributes the more they are investigated . . . will see that the course proposed does not imply a degradation of the so-called higher, but an elevation of the so-called lower."[2] From the infant's primer with its strokes and pothooks up to the pages of Newton and Spencer, we discern the same evolving aggregate, not progressively integrating simply, but simultaneously undergoing various secondary redistributions: the structural complexities thus emerging being ever accompanied by the functional complexities known as grammatical, logical, literary, scientific, and so forth. Given the indestructibility of ink and the persistence of paper, together with the various derivative laws that are their corollaries and consequences, and it can be shewn — adapting the words of our great evolutionist — not only how the grammatical elements exhibit the traits they do, but how books are evolved, thoughts generated, and civilisations achieved. But deny our fundamental datum, or, as Mr. Spencer says: "Let idealism be true, and evolution is a dream!"

[1] *First Principles*, p. 558. [2] *o.c.*, p. 556.

Very ridiculous, of course, but not more essentially ridiculous than Mr. Spencer's procedure. The plausibility of his cosmic philosophy is due entirely to the ingenuity with which he has devised a set of formularies that seem, till closely scrutinised, to carry always the same meaning; though at one time they are used in a *strictly* mechanical sense, while at another they are only figuratively mechanical. The illusoriness is the more complete and captivating because it is the ingrained habit of human intelligence to betake itself to metaphor and parable. The current scientific terminology is full of such, and we only realise that we have been talking in similes when the progress of knowledge has enabled us to outgrow them. Thus we now repudiate as fanciful the powers of Love and Hate working between the elements, as Empedocles represented; though we still talk with little misgiving of attractive and repulsive forces, of chemical affinities and bonds; speak of organisms acquiring and bequeathing, and of seeds or eggs as inheriting; and so forth. All this plenitude of metaphor is grist to the Spencerian mill. Moreover, to the 'pseudo-thinking'—I borrow his favourite phrase—which science allows to pass as sterling coin, this latest Paracelsus has added abundance of his own counterfeit.

The lesson which our reflections on the mechanical theory seemed to teach has apparently never dawned upon him, although perhaps that lesson is nowhere more impressively taught than it is in his own *First Principles*. According to that, philosophy must start from the unknowable, science from the imperceptible. Knowledge is to be unified by ruthlessly abstracting from the concrete real all qualitative specification. Celestial bodies, organisms, societies, are to be reduced to their lowest terms, viz., Matter, Motion, Force; and are to find their rationale in the instability of the homogeneous, the segregation of the heterogeneous, and the tendency of all things towards equilibrium. Surely this is not very unlike trying to find the meaning of a book by first distributing the type and then mincing them up into strokes and dots. Like the physicists who think to attain "a knowledge of what actually goes on behind what we see and feel," by treating the ideal abstractions of pure mechanics as the real things, so Mr. Spencer essays to find the fullest meaning of evolution among its emptiest symbols, to deduce the form and life of the universe from an Indeterminate and Unchanging Non-relative, which "the imbecilities of our understanding," as he tells us, prevent us from either comprehending or rejecting. The farthest point to which Philosophy, or knowledge of the highest degree of generality, can attain in seeking to comprehend this inscrutable fetish, supposed to be the Supreme Reality, is reached when all separate truths are resolved into implications of one *a priori* truth, the Persistence of Force. The experience of force is assumed to last out through the process of abstract analysis and generalisation, and to remain as long as any content remains; beyond this, we have only indeterminate, non-relative Existence or Persistence, being without content, as the supreme, ineffable generalisation of all. Thus Mr. Spencer outvies your speculative physicists in both directions; his ultimate analysis goes beyond theirs, and in his subsequent synthesis phenomena of all kinds are to be included. And by so much as the range of his formulæ exceeds theirs, by so much are his results illusory and worthless. Lord Kelvin's speculations, for example, were restricted to the deduction of material phenomena from the motions of a structureless primordial fluid; and he is careful to say "that the beginning and the maintenance of life on the earth is absolutely and infinitely beyond the range of all sound speculation in dynamical science."[1] Lord Kelvin, too, it will be remembered, proposed to test all his hypotheses by the construction, real or imaginary, of a mechanical model — thus shewing unmistakably that Matter, Motion, and Force were to be taken in a strictly literal sense. And this, of course, is, if anything, still more true of physicists of the Kirchhoff school, for whom these conceptions are pure mathematical abstractions, not real existences. How, then, does it come about that Mr. Spencer imagines he can set forth the entire history and rationale of the universe in such terms? Do mechanical models of organisms and societies arise and work before his philosophic eye, or can his transcendent mathematical genius apply the equations of motion to such phenomena and sum them up in generalised coördinates as yet undreamed of? Nothing of the sort. It is simply the superiority of ignorance that enables him to soar even in a vacuum. Severe as is the following characteristic of Mr. Spencer's powers, it is, to my thinking, as just as it is discriminating. I quote again from a review which, though anonymous, is known to have been written by a dis-

[1] *Properties of Matter*, p. 415.

tinguished lawyer and mathematician: "The flexibility of meaning that characterises well-known formulæ when they come into his [Mr. Spencer's] hands, combined with an incapacity for distinguishing between real and apparent analogies, enables him ever to find, on the one hand a principle, and on the other a multitude of examples, to support each of his positions, and imparts to his style 'the semblance of perpetually hitting the right nail on the head without the reality.' If there be any part of science that Mr. Spencer knows thoroughly, and where his positions are right ones, his writings will there be highly valuable and suggestive. But what these parts are we must learn from others, for Mr. Spencer cannot tell when he does not understand a subject; and his mind is such that it allows him to frame inductive and deductive proofs of his propositions, with almost equal facility, whether they be true or false."[1]

To pass to particulars. The hopeless vagueness of Mr. Spencer's conception of Force is notorious, and has been already sufficiently referred to. But there is a further point, which I should like to make clearer, in which Mr. Spencer is more or less at one with those whom we may call the realistic physicists as distinguished from physicists of the Kirchhoff school, — and that is in confusing abstraction with analysis. It was to such a confusion that we attributed the notion of the realistic physicist that the way to a knowledge of what actually goes on behind what we see and feel lies through hypothetical constructions in the region of abstract mechanics. Sharing in this view and unencum-

[1] *British Quarterly Review*, vol. lviii, p. 504.

bered with precise knowledge, Mr. Spencer thinks he can succeed in interpreting the detailed phenomena of Life and Mind and Society in terms of Matter, Motion, and Force. The avowed presupposition of such a synthesis is the belief that by a prior analysis those phenomena have been reduced to these lowest terms. This belief, then, I contend, is due to a confusion between abstraction and analysis.

No doubt these two processes are intimately connected, inasmuch as in abstracting we also analyse and in analysing we also abstract. And yet there is an important difference, and it is this that Mr. Spencer and others beside him have overlooked. As to the procedure in abstraction as such, what Bentham styled "the matchless beauty of the Ramean tree" has, since the days of Porphyry, furnished its classic type. Here, as every one knows, we ascend by successively ignoring essential characters. Starting from some given concrete reality, we rise through a strictly indefinite series of intermediate species or genera to the *summum genus* or *genus generalissimum*, BEING; to a conception, that is to say, devoid of assignable content and only formally distinguishable from its contradictory Non-being. As to analysis—this unfortunately is an ambiguous term. Perhaps the usage in chemistry is the most appropriate, as it is the most literal. Here then we resolve a whole into its constituent elements. And here, in contrast to abstraction, the farther we proceed the more numerous the constituents become. I assume, let me say, that among these constituents we include all those relations of what we may call the mere elements concerned, without which their subsequent synthesis would be impossible, — relations on which, quite as much as on the mere elements themselves, the nature of the real whole depends. Adopting an illustration of Condillac's, — who compared analysis to the act of taking a watch to pieces, and synthesis to that of putting it together again, — I should say the analysis was incomplete till it sufficed to insure this reconstructive process. Now when the physicist regards things from the mechanical level, we have both abstraction and analysis and also synthesis. We have abstraction in that everything — to requote Maxwell — "is considered under no other aspect than as that which can have its motion changed by the action of force." We have analysis in as far as this conception of mechanism is found to involve the three simple and independent elements of mass and space and time; and we have a basis for synthesis in the laws of motion expressing the relations of these elements. But synthesise as much as we may, the entire result remains abstract; for we cannot by synthesis introduce new elements, any more than by combining two chemical elements we can produce a compound of three. It is because they see this clearly that physicists of the Kirchhoff school repudiate the notion of attaining by merely mechanical investigations to any presentment of "what actually goes on"; and it is because he does not see it at all that Mr. Spencer must rank either as a materialist — and this he disclaims — or as a 'pseudo-thinker.'

In his so-called ultimate analysis, from which his so-called rational synthesis is supposed to build up, we have *only* abstraction, nothing left to analyse and no basis for synthesis. Let us recall some of his descriptions. How can we analyse 'the uncognisable,' that which is 'deeper than definite cognition,'[1] which "is not the abstract of any one group of thoughts, ideas, or conceptions, but is the abstract of *all* thoughts, ideas, or conceptions, that which is common to them all and cannot be got rid of, 'what we predicate by the word existence,' 'being apart from its appearances?'[2] In short, Mr. Spencer's own words shew unmistakably that his ultimate analysis is that *ne plus ultra* of abstraction, the logically unattainable apex of the Porphyrean tree, a height of abstraction from which there is no return. This abstract analytic procedure Hegel has quaintly compared to the process of peeling off the coats of an onion; now, in what Mr. Spencer calls ultimate analysis, all the coats are gone. If we are now to brush all these aside, it does not greatly matter whether we call what is left Non-being or "being apart from all appearances." It is a question of taste; some prefer one, some the other. The way back to rational synthesis is alike impossible from either. The feats by which Mr. Spencer seems to accomplish it we have admired already. From the persistence of existence to the conservation of energy and from the conservation of energy to the entire body of mechanical principles, two easy steps for Mr. Spencer, and he is in line with the mechanical theory. Having thus conjured himself back from a height of abstraction, avowedly devoid of all definite content, to a definite content admitting of analysis,

[1] *First Principles*, p. 192. [2] *o. c.*, p. 95.

we are not surprised to find Mr. Spencer skilful enough to make a show of building up the whole fabric and essential history of life and mind and society in terms of that content, *i.e.* in terms of Matter, Motion, and Force. Having advanced from the indefinite residuum as far as these three coats of his onion and their laws, it seems no longer an impossible feat to conjure all the rest out of these. But I contend that it is only conjuring. The new elements are adroitly taken up as the synthesis advances, although they seem to have been swept from the board before the performance commenced. The process is not legitimate because they are not avowed as parts of the ultimate analysis; because, in fact, this supposed analysis is incomplete, is not analysis but abstraction, on the way to which these elements were left entirely aside.

Mr. Spencer's remarkable qualifications for this kind of work I have tried already to describe and to illustrate —perhaps at undue length. But there is one characteristic of evolution which lends great additional plausibility to his enterprise and to other like enterprises; I mean the extremely gradual advance, the general absence of all discontinuity, that pertains to nature's developments —that trait which is embodied in the familiar axiom, *Natura non facit saltum.* In a nebulous haze compared with the endless variety of the solar system; in the dance of drops in a fountain of water compared with the physiological processes in a living organism; in the *Amœba* compared with *Homo sapiens;* in a group of savages uttering incantations round a newly fallen meteorite compared with the Fellows of the Royal Society discussing argon, — we see the most divergent extremes of kind. Yet there are innumerable intermediate forms connecting these several extremes by insensible degrees. When we consider the extremes by themselves, as our forefathers for the most part did, the explanation of the more complex extreme confronts us as a formidable problem, however adequate our explanation of the simpler extreme may appear. But nowadays, familiarised as we are with the successional continuity of the intervening stages, we are inclined to imagine either that there is no problem at all, or that, if there is, the problem is solved. Psychologically this may be readily accounted for. Certain well-known sentences of Hume here apply exactly: "The passage is . . . so smooth and easy, that it produces little alteration on the mind, and seems like the continuation of the same action. . . . The thought slides along the succession with equal facility, as if it considered only one object; and therefore confounds the succession with the identity."[1] And so we can understand why, as Sigwart remarks, "the notion of development has sometimes been handled like a logical charm by means of which phenomena hitherto inexplicable are explained with ease." "It is," he continues, "as if we should say, that though force is required to lift a weight a given height perpendicularly, yet if the weight is placed on an inclined plane and this made very long, so that over small lengths the weight would rise only imperceptibly, it might really rise of itself; for the force diminishing as the time increases, if the time taken were very long, force could be dispensed

[1] *A Treatise of Human Nature*, Green and Grose's edition, vol. i, p. 492.

with altogether."[1] But in truth, the law of continuity does not dispense with causal laws, however much it may facilitate genetic description or aid the dissolving views of imagination. Evolution, so far from being a self-sufficient explanation of what are called its results, has itself to be explained; like other processes, it must have its adequate cause. But not merely so. Allowing science to content itself with description, as we have seen that it tends to do, still it is impossible, as we have also seen, to convert the dead letters of the mechanical alphabet into the living sense of things. Other and higher conceptions have to be employed, and no continuity or smoothness of transition will account for these; though it may enable them to slip in easily and unawares, thereby committing science to sophisms of the Sorites type, which philosophic reflection may find it hard completely to expose. In truth the topic is too difficult and would divert us too widely from our immediate theme if we attempted to discuss it fully here. My present purpose is simply to call attention to this feature of evolutionism.

In pursuance of this object I will only remark further that those serious gaps between the sciences which we have already noticed,[2] so far from being, as we might expect, a hindrance to the effective working of that logical charm seemingly pertaining to the notion of development, really enlarge its scope and enhance its potency. Take, for example, the gap between the inorganic and the organic. Of the origin of life, *if it ever did originate*, we have absolutely no knowledge. But, on the one

[1] *Logic*, § 100, 15. [2] Cf. Lecture I, pp. 8 ff.

hand, there is no definite limit to the possible complexity of mechanical processes, nor any definite limit, on the other, to the possible simplicity of life. Thus in science we have every facility and many temptations to assume that somewhere in the *terra incognita* between physics and physiology mass-aggregates become so configured as to take on the functions and individuality of organisms. Meanwhile — and again contrary to expectation — the progress of knowledge and especially of that systematic reflection concerning knowledge, which takes knowledge itself as the object of science, the science we call epistemology, instead of making this conjectural transition easier, renders it increasingly hazardous and difficult. In proof of this it may be enough here to contrast the light and airy way in which Mr. Spencer glides over this problem, with the confidence of physicists like Lord Kelvin or Helmholtz, or of physiologists like Liebig and Pasteur, that mechanical theories as to the origin and maintenance of life are hopeless.

To be sure Mr. Spencer tells us, when hard pressed by critics, that of the synthetic philosophy "two volumes are missing" — the two important volumes on Inorganic Evolution. "The closing chapter of the second of these volumes" — he continues — "*were it written*, would deal with the evolution of organic matter — the step preceding the evolution of living forms. Habitually carrying with me in thought the contents of this unwritten chapter, I have, in some cases, expressed myself as though the reader had it before him; and have thus rendered some of my statements liable to misconstruction."[1]

[1] *Principles of Biology*, vol. i, p. 480. Italics mine.

Surely this is a situation not wanting in humour — or in pathos! Who is the more to be pitied: the sympathetic readers, who — through no fault of their own, as Mr. Spencer allows — have misunderstood, lacking as they have done for thirty years these two missing volumes of the stereotyped philosophy; or poor Mr. Spencer himself, with these unwritten volumes in his teeming brain, compelled all that time to see his statements misconstrued? Still we must take facts as we find them. During the thirty years in which Mr. Spencer has been engrossed with this interpretation, a whole generation of biologists has striven hard, but striven in vain, to bring this truth to light. For all but Mr. Spencer, at any rate, the origin of life has remained a mystery.

So far as I can gather from his summary references to this unwritten section of his philosophy, Mr. Spencer's procedure there differs in no respect from his procedure generally. Unless I, too, misconstrue it, it exactly illustrates what I have said, and amply justifies the animadversions I have made. On the one hand we have statements purporting to be strictly mechanical; on the other, conceptions not mechanically intelligible slipping in unawares and gradually changing the *venue*. More definitely, on the one hand we have a chemical molecule increasing in complexity till we reach the proteids. Then — I here quote Mr. Spencer — "the supposition (justified by analogies)" that atoms of sulphur or phosphorus "may be a bond of union between half a dozen different isomeric forms of protein." And so, — continues Mr. Spencer, and getting bolder, — "a moment's thought will show that, setting out with the thousand isomeric forms of protein, this makes possible a number of their combinations almost passing the power of figures to express. . . . Molecules so produced, perhaps exceeding in size and complexity those of protein as those of protein exceed those of inorganic matter, may, I conceive, be the special units belonging to special kinds of organisms."[1] So far, except that Mr. Spencer premises that the ordinary idea of mechanical action must be greatly expanded, *i.e.* that we are to take the full benefit of mechanical hypotheses concerning physical and chemical phenomena — so far, with this proviso, we are still within the range of our lowest terms, Matter and Motion. We are only asked to imagine a very complex cluster of very complex chemical molecules. But, on the other hand, we find ourselves presently approaching this aggregate from the standpoint of biology; and we hear our oracle saying as follows: "Exposed to those innumerable modifications of conditions which the Earth's surface afforded, here in amount of light, there in amount of heat, and elsewhere in the mineral quality of its aqueous medium, this extremely changeable substance must have undergone now one, now another, of its countless metamorphoses. And to the mutual influences of its metamorphic forms under favouring conditions, we must ascribe the production of the still more composite, still more sensitive, still more variously-changeable portions of organic matter, which, in masses more minute and simpler than existing Protozoa, displayed actions verging little by little into those called vital. . . . Thus, setting out with inductions from the experience of

[1] *Principles of Biology*, vol. i, p. 486.

organic chemists at the one extreme, and with inductions from the observations of biologists at the other extreme, we are enabled deductively to bridge the interval — are enabled to conceive how organic compounds were evolved, and how, by a continuance of the process, the nascent life displayed in these became gradually more pronounced."[1] In other words, going farther in the way of complexity than chemical inductions directly warrant, and farther in the way of simplicity than biological observations directly justify, these two lines of conjecture will meet somewhere in the unknown interval, and *there* will be the source of life. After this triumphant deduction, is it not captious and unkind to object, when — without further explanation — portions of an extremely changeable stuff are declared to have assumed the unity and permanence of individuals? Or when the particles of this stuff, spoken of as living, are credited with "an innate tendency to arrange themselves into the shape of the organism to which they belong,"[2] 'tendencies derived from the inherited effects of environing actions?' Or again when, though scornfully repudiating the hypothesis of a *nisus formativus*, or vital principle, Mr. Spencer allows himself to talk of "the polarities of the molecules determining the *direction* in which the *power* [of environing forces] is turned?"[3]

Instead of pausing to comment, let us rather take one more sample of Mr. Spencer's procedure, which lies on the way to our next topic — the transition from life to mind. "The broadest and most complete definition of

[1] *Principles of Biology*, vol. i, pp. 483 f.

[2] *o.c.*, vol. i, pp. 180 f. [3] *o.c.*, vol. i, p. 488.

Life," he tells us, "will be *The continuous adjustment of internal relations to external relations.*"[1] This we are to understand as a dynamic statement, and possibly in the instance first given to exemplify it we might contrive so to understand it — the instance being the correspondence between food assimilated and the temperature of the environment. But how are we to find a dynamic statement in such an instance as this: "A sound or a scent wafted to it on the breeze prompts the stag to dart away from the deerstalker"? A child would understand that adjustment here does not mean any "transformation or equivalence of forces," and that when the stag halts panting in a corrie five miles off, the internal change from fright to a sense of security cannot, like the external change, be exhibited by geometrical or dynamical diagrams. Yet Mr. Spencer's "broadest and most complete definition" is meant to cover both these cases; spite of the important difference that in the one 'internal relations' refer to states of the organism, and involve all the three physical terms, space, time, and mass; while in the other 'internal relations' refer to states of mind, and so far can involve neither space nor mass. Now we shall all admit that it is a somewhat hazardous enterprise to set out "to interpret in terms of Matter, Motion, and Force" — such, it will be remembered, is the classic phrase — phenomena into which it is allowed that matter, motion, and force do not enter. The difficulty is two-fold: first, to get rid of extension; and then, since with extension matter goes too, to get back the real in some

[1] *Principles of Biology*, vol. i, p. 80.

other form. But it is just in these 'disastrous chances' that Mr. Spencer's characteristics come out. That you may learn in his own words how he resolves the first difficulty, how from internal relations of the organism he passes over to internal relations of the mind, let me quote from his *Principles of Psychology*. The following is part of a chapter devoted to elucidating the nature of intelligence:—"The skin, then, being the part immediately subject to the various kinds of external stimuli, necessarily becomes the part in which psychical changes are originated. . . . Speaking generally, therefore, we may say that while the physical changes are being everywhere initiated throughout a *solid*, the psychical ones, or rather those out of which psychical ones arise, admit of being initiated only on a *surface*."[1] So one dimension of this too, too solid flesh melts; to understand how the other two disappear let us hear Mr. Spencer further. "Those abilities which an intelligent creature possesses, of recognising diverse external objects and of adjusting its actions to composite phenomena of various kinds, imply a power of combining many separate impressions. These separate impressions are received by the senses—by different parts of the body. If they go no further than the places at which they are received, they are useless. . . . That an effectual adjustment may be made, they must all be brought into relation with one another. But this implies some centre common to them all through which they can pass; and as they cannot pass through it simultaneously they must pass in suc-

[1] *Principles of Psychology*, vol. i, p. 401.

cession, so that as the external phenomena responded to become greater in number and more complicated in kind, the variety and rapidity of the changes to which this common *centre*[1] of communication is subject must increase — there must result an unbroken series of these changes — there must arise a consciousness."[2] Just as extension reduces to a point, consciousness appears!

It would look as if a punctual seat of the soul were as much a necessity for Mr. Spencer as it was for Descartes. But Mr. Spencer's dynamic principle recognises no substance but matter, and that has gone with space. This brings us to the second difficulty.

How are we to interpret the intelligent creature for whom this hurrying single file of impressions is brought into relation? Since it cannot be what it ought to be (if it is to be rationally built up according to Mr. Spencer's ultimate analysis), since it cannot be matter, and must be something, what, we wonder, is it? Now for the *deus ex machina*. Turning to his chapter on the Substance of Mind, we read: ". . . The concept we form to ourselves of Matter is but the symbol of some form of Power absolutely and forever unknown to us; and a symbol which we cannot suppose to be like the reality without involving ourselves in contradictions. . . . Also the representation of all objective activities in terms of Motion, is but a representation of them and not a knowledge of them. When with these conclusions . . . we join the conclusion lately reached that Mind also is unknowable, and that the simplest form under which we can think of its substance is but

[1] Italics mine. [2] *Principles of Psychology*, vol. i, p. 403.

a symbol of something that can never be rendered into thought; we see that the whole question is at last nothing more than the question whether these symbols should be expressed in terms of those, or those in terms of these—a question scarcely worth deciding."[1]

What's in a name? The rose by any other name would smell as sweet, and when it is no longer convenient to call our 'real' matter, why not call it mind? Why not indeed? Most of us here, I dare say, have no objection. Still the somersault is a little startling even from our poet-philosopher, who in concluding his *First Principles* we remember had said: "The interpretation of *all* phenomena in terms of Matter, Motion, and Force is nothing more than the reduction of our complex symbols of thought to the simplest symbols." Our surprise is the greater because here in this chapter on the Substance of Mind he calmly remarks: "It seems easier to translate so-called Matter into so-called Spirit, than to translate so-called Spirit into so-called Matter (*which latter is, indeed, wholly impossible*). . . . Our only course," he continues, "is constantly to recognise our symbols as symbols only; and to rest content with that duality of them which our constitution necessitates."[2] But now what has become of the complete unification of the knowable in view of this utter dualism; and how now are the complex facts of intelligence and morality, of man and society, to be rationally 'built up' on the doctrine of the conservation and transformation of energy? No wonder Mr. Spencer has ever and anon to enter a *caveat* such as this, which

[1] *Principles of Psychology*, vol. i, p. 159. [2] *o.c.*, vol. i, p. 161.

occurs in his treatment of social phenomena: "Though evolution of the various products of human activity cannot be said directly to exemplify the integration of matter and the dissipation of motion, yet they exemplify it indirectly."[1] From synthetic interpretation to indirect exemplification is verily a descent, nay, is the most palpable failure. How very indirect even the exemplification is may be judged from Mr. Spencer's final statement of the psychological side of his great primordial truth, viz., that "all mental action whatever is definable *as the continuous differentiation and integration of states of consciousness.*"[2] This does not seem to *mean* the same thing as the continuous integration of matter and dissipation of motion; still it *sounds* a little like it.

Here, then, is a thinker really following where he essays to lead, professing to give the sciences their bearings, but in fact losing his own as he goes along. He looks at things, first of all, chronologically, and begins with the generalities of abstract dynamics, which he mistakes for natural laws. The gap between this abstract science and our empirical knowledge concerning physical phenomena, together with the whole group of physical sciences, is passed over. And when Mr. Spencer, omitting two whole volumes, resumes his task with what he calls the interpretation of Organic Nature, he seems quite unaware that he has passed not only from the abstract to the actual, but from the mechanical to the teleological. Regarding living things as a whole,

[1] *First Principles*, pp. 318 f.

[2] *Principles of Psychology*, vol. ii, p. 301.

we find that what is clearest about the lowest forms is organization, and what is clearest in the highest is mind. Midway then — there is a transition point in the evolutional drama where the poet glides easily over from the physical standpoint to the psychical, still, however, dealing with the facts chronologically. Then suddenly he ceases from this forward or synthetic movement, and at the close of his psychology sweeps back analytically, and, like a mighty boomerang, demolishes his first starting-point. In place of it there arises finally what is poetically styled "Transfigured Realism,"[1] a final tableau wherein every philosophy, from Scepticism up to Absolute Idealism, finds something to be thankful for and is anon swallowed up.

[1] See *Principles of Psychology*, pt. vii, General Analysis, last chapter.

LECTURE X

BIOLOGICAL EVOLUTION

The Lamarckian, Darwinian, and ultra-Darwinian theories generally compared. Natural selection by itself non-teleological. Attempts to assimilate the biological with the physical. Two difficulties in the way. These lead to the question: Is there not a teleological factor operative throughout biological evolution?

Teleological and non-teleological factors distinguished. Darwin recognised both. Only so far as both are present has 'struggle for existence' any meaning. The question raised equivalent to inquiring how far mind is concomitant with life. Naturalism confident that life is the wider conception, and appeals to the facts of plant-life. 'Continuity' seems to help it, but really works both ways. The case argued. The levelling-up method the simpler. Objections to this considered: (1) *Reflexes;* (2) *The character of plants again. Recent views on this point.*

Restatement of the position reached. Antagonism of organism and environment: the latter, then, not the source of life. 'Vital force' unworkable. Turning to the facts of mind we have: (1) *Self-conservation;* (2) *Subjective selection. The meaning and significance of these. Their distinctness from, and relation to, natural selection.*

IN passing, as we do in this lecture, to the narrower subject of biological evolution, we find no serious attempt made to account for the origin of life or to reduce the facts of life to those of a mechanism. The problem here is merely to explain the diversity of living forms, and that not by the help of mechanical, but of biological, conceptions. The origin of species by descent from some primitive form is assumed as the starting-point.

Then we have two widely different, but not incompatible, theories, — that of Lamarck and that of C. Darwin — to shew how, as the latter puts it, "whilst this planet has gone cycling on according to the fixed law of gravity, from so simple a beginning endless forms most beautiful and most wonderful have been, and are being evolved."[1] The doctrine of *special* creation is, by common consent, disallowed as unscientific. This of course leaves the general question of creation untouched. Still, as respects teleological conceptions, the two dominant theories of biological evolution are by no means on the same footing. The extreme Darwinian theory, as held, for example, by Wallace or Weismann, but strongly discountenanced by Darwin himself, seems — if pressed to its logical consequences — to leave but scant space for any notions of purpose or end.[2] Natural selection works blindly upon promiscuous variations blindly produced. So mechanical is the whole *milieu* that repeated attempts have been made to extend the range of natural selection, so understood, to the evolution of stellar systems, chemical elements, and the like. Such an extension would be impossible with the Lamarckian theory, as the mere citation of the second of the four laws given in the *Histoire Naturelle* will shew: "The production of a new organ in an animal body results from a new want arising and continuing to be felt, and from the new movement which this want initiates and sustains."[3] According to Lamarck, then, variations are

[1] *Origin of Species*, sixth edition, last sentence.

[2] Cf. Romanes, *Darwin and After Darwin*, vol. ii, ch. 1.

[3] *o.c.*, edition 1815, t. i, p. 181.

due to a psychical factor; but for the theory of natural selection it is immaterial how they are produced. Given the indefinite production of varying individuals, and given also restriction in the number that can simultaneously exist, and it is obvious that some individuals must be excluded and disappear; if for no other reason, at any rate, for want of standing-room. Unless the selection is a pure affair of chance, the variations themselves must determine it: in one case — the question being one of standing-room say — the highest specific gravity, in another the lowest, might constitute the requisite fitness. So in economic exchange, wherever supply exceeds demand, such principles of selection come into play, and with one commodity cheapness is the ground of fitness, with another taste, with another novelty, with another utility in the narrower sense, and so on. Such instances bring out still further the difference between the Lamarckian and the Darwinian, or more correctly the ultra-Darwinian standpoint. For Lamarck, the fitness must relate primarily and essentially to the competing individual; for Wallace or Weismann it might primarily and essentially relate to the selecting agency. Thus in sorting shot those pellets are selected that roll down an incline quickest; in sorting emery powder those particles are selected that take longest to sink in water. In short, for the ultra-Darwinian view, life need imply no more than the indefinite production of varying individuals. Struggle for Existence here becomes simply a figure of speech, not the stern reality first depicted by Malthus, to whom, I believe, the phrase is due. In the *Origin of Species* Darwin himself calls attention to this: "I should premise," he says, "that I use this term in a large and metaphorical sense."

A similar remark applies to the phrase Natural Selection. As to this let me quote from a letter of Wallace to Darwin (*Life*, ii, p. 46). He writes: "The term 'survival of the fittest' is the plain expression of the fact; 'natural selection' is a metaphorical expression of it, and to a certain degree indirect and incorrect, since . . . Nature . . . does not so much select special varieties as exterminate the most unfavourable ones." But even 'survival of the fittest' is *not* a plain expression of what logically follows from the ultra-Darwinian premisses. The notion of fitness is used just as metaphorically as that of struggle or selection, for fitness is in strict propriety a teleological conception, and there is nothing teleological in those premisses. There is only what Mr. Spencer would call equilibration: neither struggle for life, nor selection by nature, nor survival of the best, but simply conservation of the stablest; as in a mass of chemical elements capable of combining, compositions, double decompositions, neutralisations, expulsions go on, stronger affinities and avidities overcoming weaker, till the stablest and most permanent combinations are reached.

The mechanical theory of evolution, indeed, is, as we have seen, bent on assimilating the biological to the chemical in some such fashion. But in the way there are two difficulties. In the first place, if we look broadly at the world of living things and compare it with the inanimate world, we are at once confronted by a striking difference. In the latter we note a general downward trend, the resolution of potential energy into kinetic, and then of available forms of this into unavailable; in other words, we find a uniform tendency to pass in the shortest and easiest way to physical quiescence, fixity, and equilibrium. But in the organic world, on the contrary, we find a steadily increasing differentiation of structure and composition, entailing a large storage of potential energy. We see this as we advance from plants to animals, from invertebrate to vertebrate, from cold-blooded vertebrates to warm-blooded, from brutes to man. And if we take into account what may be regarded as the by-products of living things, — their stores of food, the snares they make, the habitations they build, — the same characteristics are still present, notably so, of course, in the products of human skill. The inorganic world has nothing to match dynamite, Liebig's Extract, a steam-engine, or a ship-torpedo. It is impossible in the present state of our knowledge to bring such results under the *facilis descensus* principle of least resistance, which dispenses with all conception of guidance and direction, and can give no meaning to adaptation, fitness, or worth. And, as has been urged in earlier lectures, it seems absurd to attempt ever to refer those results to such a source, unless they can at the same time be regarded as rare and exceptional manifestations of that principle when working on a very vast scale.

The second of the difficulties mentioned runs parallel to the first; it is, in fact, this advancing complexity regarded psychologically. Here we are only sure of the latest term of the series; how the earliest terms are constituted we can only vaguely guess. In the case of man and the higher animals, there is no doubt that the instinct of self-preservation and the struggle for existence are realities; no doubt, that needs and wants lead to movements; or that improvement comes only by repetition and effort, that practice makes perfect. The only doubt is whether what is thus acquired in one generation becomes in any measure the inborn heritage of the next; but with this burning question we are not for the moment concerned. We have only to demand recognition of the truth that in this advancing psychical complexity, at any rate, the teleological character of the facts is unmistakable; no other conception is adequate. Thus there arises this question which is for us the important one: Is not this teleological factor operative throughout the whole range of biological evolution at least; so far, that is, as we find the downhill trend distinctive of the inanimate world to be counterworked?

As a preliminary to the discussion of this question, it will be well to define a little more exactly what is to be understood by the phrase 'teleological factor,' and to distinguish it from the other factors implied. If Lamarck had happened to ask himself: How the leopard came by its spots, as well as how the giraffe acquired its long neck, it is very unlikely that he would have ventured to give the same explanation of both. Continued use in stretching might have enabled the giraffe to add a cubit to his stature, a continued use to which the need of food might lead; but use or need could hardly help the leopard to change its skin, even though the change should facilitate the capture of its prey. A more probable explanation here is the purely Darwinian one, that skin-colouration being specially liable to vary, a variation simulating the play of sunshine through foliage had favoured the ancestors of the leopard when lying in wait to pounce upon their spoil; and that such variation had been perfected by natural selection. At any rate, though not forgetting much striking evidence of a functional and more or less voluntary connexion between an animal's colour and its immediate surroundings, we may fairly take the leopard's spots, the tiger's stripes, or the lion's tawny hue, as instances of fortuitous or non-teleological[1] adaptation. Another factor that may be classed as non-teleological, though it is one of minor importance, is that described by Darwin as "the direct action of external conditions," such as climate and food. This is the factor on which Buffon laid stress, and to which Buckle and the materialists are fond of appealing, an appeal culminating in the *mot* of Moleschott, *Der Mensch ist was er isst.* In contrast to these factors of biological evolution, then, the meaning of what I have proposed to call the teleological factors will become clearer. Among these I think we might enumerate three. First, the Lamarckian principle already referred to, secondly, Darwin's Sexual Selection, and lastly, Human Selection, on which Wallace has the merit of laying especial stress.[2]

[1] Non-teleological, that is, within the range of strictly biological ideas.

[2] I refer, of course, to his contention that the moral and intellectual nature of man cannot be explained by natural selection. See his *Darwinism*, ch. xv.

The name of Lamarck has been so long in disrepute that it would be rash to cite any theory of his, if there were not at length among biologists a manifest reversion in his favour. Opposed to the neo-Darwinians who profess to see in natural selection far more than ever Darwin publicly[1] claimed for it, there is also a numerous neo-Lamarckian school, who replace the fanciful illustrations that served to discredit Lamarck's speculation by an imposing array of facts in its support. Such materials were not in existence in Lamarck's day; and from the free use of what material there was, he seems to have been cut off, partly by blindness and partly by poverty. It was thus easy for Cuvier, that master of details, to turn the laugh against poor Lamarck, and as the favourite of Napoleon, to use his political influence against "the transformists," as the Lamarckians were called.[2] So it came about that when Darwin was working out his *Origin of Species*, Lamarck's doctrines were in general discredit, and yet had never received an impartial hearing. Darwin's letters shew his anxiety lest these doctrines should be identified with his own. "Heaven forfend me," he wrote to Hooker in 1844, "from Lamarckian nonsense of a 'tendency to progression,'[3] 'adaptations from the slow willing of animals, etc.' But the conclusions I am led to are not widely different from his; though the views of change are wholly so." Nevertheless, as time went

[1] Cf. Osborn, *From the Greeks to Darwin : an Outline of the Development of the Evolution Idea*, 1895, p. 236.

[2] Cf. Osborn, *o.c.*, p. 196.

[3] Which, by the way, it would seem Lamarck did not hold. Cf. Osborn, *o.c.*, p. 237.

on, Darwin was led by his own further studies and observations to include the Lamarckian factor among his 'views of change.' As Romanes says: The longer he (Darwin) lived . . . the less exclusive was the *rôle* which he assigned to natural selection, and the more importance did he attribute to the supplementary factors." Thus, to quote one instance: in the conclusion to his last edition of the *Origin*, Darwin protests against those who have misrepresented him as attributing the modification of species exclusively to natural selection, and expressly refers to it as "aided in an important manner by the inherited effects of the use and disuse of parts,"[1] *i.e.* by what is commonly called the Lamarckian factor. There is then after all no imprudence in citing this principle.

In calling this factor teleological there is, of course, no intention of connecting it with the old view that each species was immediately designed and directly fashioned to occupy a fixed place in a supposed 'plan of creation.' As already said, Lamarck, equally with Darwin, assumed the evolution of all species from a common source. I call this factor teleological, simply, then, on the ground that it presupposes conscious, or at least sentient, activity directed to the satisfaction of needs, appetites, or desires; psychical activity, in a word, as distinct from physical passivity and inertness. It implies an impulse to self-maintenance and betterment, which so far become ends. Only so far as such conceptions are applicable, is there any meaning in talking of struggles to survive, or in saying, as Darwin

[1] *Origin of Species*, sixth edition, p. 421.

does, that "Natural selection acts solely by and for the good of each."[1] Sexual selection, and still more obviously human selection, can be brought under the same head, and call here for no further notice.

And now we may take up again the question: Is this same teleological factor operative throughout the whole range of biological evolution, or is it confined to those higher forms of life which have some obvious resemblance to our own? The question is one that seems to have important bearings on our main inquiry, as I shall hope to shew later on. Broadly put, the question is, How far is mind concomitant with life? With this question neither Lamarck nor Darwin has dealt explicitly; in fact biologists as such, for the most part, ignore it. But naturalism, of course, confidently assumes that life is the wider conception, that mind is but an occasional accompaniment of organisation and is certainly never a cause of it; just as it confidently assumes organisation to be but a special arrangement of inert masses and the effect of mechanical forces. Perhaps, however, on closer inspection, life, so regarded, may prove as insoluble a riddle as mind, so regarded, is likely to prove. Comparing the lower forms of life with the higher, it is at once obvious that the non-teleological factors seem more exclusively the efficient ones the lower down the scale we go, while the teleological factors come more clearly into play the higher we ascend. It is true that even plants respire, imbibe, and assimilate; and that among all but the lowest, as among all but the lowest animals, there are

[1] *Ibid.*, p. 162.

differences of sex. "Still," it will be replied, "only poets talk of 'the loves of the plants'; science has no ground for ascribing to them activities determined by hunger and thirst, or other organic needs. And yet how impressively diverse and complex are the developments to which, by the operation of the non-teleological factors, the vegetable kingdom has attained. The apparatus by which the bee orchis or the garden sage secures the aid of insects in its fertilisation, or that by which the crane's-bill or the thistle scatters its seed, exceed in ingenuity the snares of the spider or the ant-lion, are comparable indeed even with human devices like the parachute or the sling. Such instances, too, it must be remembered, are not the exception, but the rule, in the economy of plants; whole libraries might be devoted to the description of them. Such marvels of organisation, "it is argued," has natural selection accomplished by steadily eliminating unpropitious variations, entirely unaided by any sort of spontaneous impulse, sentient preference, or organic memory, — to say nothing of conceptions so mystical as the entelechies[1] of Aristotle, the *nisus formativus* of later writers, and other notions equally transcendental. If, then, nature alone can advance thus far before psychical phenomena appear at all, why suppose, when these are present, that they are more than concomitant, why attribute to them any share in the organising processes? At every step in the genealogical descent, alike of plant and animal, the germ is built anew into the parental form by a like inevitable process:

[1] The mysticism now commonly associated with this conception seems mainly due to the neo-Platonists and the Scholastics.

the acorn is here not more passive than the egg; in each alike the embryo recapitulates the stages by which it has been evolved. Why then suppose psychical factors to be necessary to the one result, when they are dispensed with in the other? It is much like saying that though the coiled spring works the meat-jack, we must suppose a musical box to be worked by the tune it plays."

Such language, I think, fairly represents the levelling-down method to which naturalism is led. For this method it claims the advantages of clearness and simplicity; on the ground, as urged by Huxley, that by thus extending the range of matter and law, it is enabled to substitute the verifiable for the unverifiable, to replace, by a single objective standpoint, subjective standpoints that may be innumerable. To the psychophysical doctrines in which it culminates, I shall hope to invite attention six months hence. In common with other attempts to make lower categories take the place of higher ones — striking instances of which we have discovered in the exposition of Mr. Spencer — this procedure gains greatly in verisimilitude by the use it can make of the principle of continuity, that cardinal principle of all theories of evolution. But it should not be forgotten that on the levelling-up method the principle of continuity is equally available. The scale of life is just as continuous from Man to the *Protista* as it is from the *Protista* to Man. To understand human actions we have to take account of mind; on the one method, then, we carry back this conception of mental determination, our teleological factor in other words, as far as we can. In so doing, we may claim to be describing the unknown in terms of the known. Imagination, it is true, will not enable us to depict what Huxley would call the *psychoses* of creatures so far beneath us. But that alone does not invalidate the conception; if it did, a good many scientific ideas would become illegitimate. On the other hand, the levelling-down method has always more or less definite pictures to offer of the structure and movements, as also of the phylogeny and the ontogeny of each new member in any series of living forms, as it follows forward the continuous interaction of variants and environment. But then comes the difficulty, which led us first of all to inquire whether teleological factors were not throughout indispensable.

Now, even if we were to grant the theory of psychophysical parallelism, this alone would not justify us in saying that life is a wider fact than mind. Simple forms of life might have as concomitants equally simple forms of mind. We have allowed that the psychologist is here at a disadvantage just as the biologist, or rather the physiologist, is correspondingly at a disadvantage, at the opposite extreme. We cannot certainly discern or imagine the mental states of creatures whose entire organism consists of a single cell. But even the biologist in such a case is found to infer much greater complexity of structure than ever the microscope will enable him to see. The psychologist is equally entitled to infer the presence of appropriate mental concomitants in these unicellular organisms, if the facts of life as a whole are made clearer by so doing. I have only time to deal here with such general considerations, but, in truth, the more the protoplasmic movements, even of the lowest plants, are studied, the more they are found to resemble actions determined by stimuli and to deviate from the mechanical motions of inert masses. To such studies we owe in large measure what its opponents regard as a recrudescence of superstition, and its upholders call '*neo-vitalism.*' However, without discussing detailed observations, the serious difficulty just now mentioned as besetting the levelling-down method is — to say the least — greatly simplified by the opposite method, which assumes that mind is everywhere coincident with life. That tendency to disturb existing equilibria, to reverse the dissipative processes which prevail throughout the inanimate world, to store and build up where they are ever scattering and pulling down; the tendency to conserve individual existence against antagonistic forces, to grow and to progress, not inertly taking the easiest way, but seemingly striving for the best, retaining every vantage secured and working for new ones, — this complex characteristic of all forms of life belongs also to mind. Correlated with mind these characteristics are intelligible; but to interpret them literally in terms of physical interaction, and apart from mind, is surely impossible. However we resolve the problem as to the connection of mind and matter, it is then, we may conclude, unquestionably a simplification to infer that wherever a material system is organised for self-maintenance, growth, and reproduction, as an individual in touch with an environment, that system has a psychical as well as a material aspect.

There is one very plausible doctrine not uncommon among psychologists and countenanced, as we should expect, by Mr. Herbert Spencer, that stands in the way of this view. Mr. Spencer, as we have seen, imagines consciousness to arise when physiological processes become too complex to work automatically. Up to that point the reactions of the organism are simply reflexes, beyond it they are volitions: and since we are usually unconscious of reflex movements, and since, moreover, they are usually beyond our control, it is concluded that reflexes only indicate life but do *not* implicate mind. But looked at more closely, this conclusion is at variance with the principle of continuity, that fundamental axiom of evolutional theory; and it is besides, as I have urged at length elsewhere,[1] not really borne out by empirical psychology. Reflex movements are called mechanical or automatic, because of the uniformity, promptness, and precision with which they occur. None the less, even the simplest of them depend on the exact adjustment of structures often very complicated. Accordingly the biologist makes large drafts on time and appeals freely to natural selection to account for their ultimate perfection. But during all this time the various more or less abortive attempts thus leading up to an eventual automatic regularity ought, on Mr. Spencer's theory, to be accompanied by consciousness. Moreover, when we turn to our own experience, this is precisely what we find in all those cases where long practice makes perfect, and where feats of dexterity and the like become, as psychologists say, secondarily-automatic.

Another seeming hindrance to the view I am attempting

[1] *Encyclopædia Britannica*, article *Psychology*, pp. 42 f.

to propound and defend, is the one I was just now referring to, viz. the character of plants. But strangely enough this difficulty has been in the main removed by the biologists themselves. For it would hardly be going too far to say that Aristotle's conception of a plant-soul, though it would be expressed in other language, is tenable even to-day, at least as tenable as any such notion can be at a time when souls are out of fashion. The popular idea of the three natural kingdoms, mineral, vegetable, animal — plants developing from minerals, and animals from plants, as represented by the ingenious device on the covers of Mr. Spencer's volumes — has been long abandoned. If such tripartite division is retained at all, the animal it would seem should rather precede than follow the plant. For the earliest stages of plant development resemble those of animal development, though according to all the rules of evolutional propriety the converse would hold, if plants were first in order. But modern biology, as I understand, assigns the first place in the organic world to a kingdom of *Protista*, living things, that is, from which individuals with the definite characteristics of plants and animals were afterwards differentiated. The *Protista* display in a marked degree the motility and sensibility specially associated with animal life. Certain of these freely-moving creatures are supposed to have assumed a sessile position on the earth, and so to have become plants or earth-parasites, as such developing their capacity to build up protoplasm direct from its mineral constituents, but degenerating in respect of their distinctively animal traits, in consequence of their fixity of habitat. The distinctively animal kingdom, on the other hand, it is conjectured, began with the first protist, who anticipated by untold ages the feat of little Jack Horner, and did what animals have been doing ever since — appropriated and devoured the ready-made protoplasm. "The easy nutrition which ensued," says Professor Cope, "was probably pleasurable, and once enjoyed was repeated and soon became a habit. The excess of energy thus saved from the laborious process of making protoplasm was available as the vehicle of consciousness and motion."[1] But all such conjectures aside — it is at any rate certain that plant protoplasm and animal protoplasm are essentially one and the same; that the animal functions of motility and sensibility pertain to all protoplasm as truly as the vegetable function of assimilation and reproduction; that from unicellular organisms, the *Protista*, leading the free-swimming life of animalcules and yet endowed with the plant's power of transforming inorganic matter, there arose both unicellular organisms, the *Protozoa*, retaining and developing the former characteristics; and also unicellular organisms, the *Protophyta*, with the antithetic traits; and finally that from the *Protozoa* and *Protophyta* respectively all the more complex animal and vegetable organisms have been evolved.

Let me now try by way of recapitulation to explain in what sense I understand mind in thus concluding that it is always implicated in life, or that, in other words, a teleological factor, analogous to that of Lamarck, is operative and essential throughout all biological evolution. Let us begin with the opposition of the living individual, or organism, and its environment. These

[1] *Primary Factors of Organic Evolution*, 1896, p. 514.

terms are, in biology, strictly correlative, just as in psychology the terms subject and object are. This correlation is one that only appears with life; the physicist never gets beyond the action and reaction of bodies that are not properly individuals. On looking at this relation of organism and environment more closely, we discover that it is essentially an antagonism. Whether living or dead, the organism is equally a material system, and its *death* makes no change in what we may call the attitude of the environment. What this attitude is, is therefore shewn by the processes that then ensue. These processes, one and all, belong to the downhill trend characteristic of inorganic changes; adopting, but somewhat extending, a convenient physiological term, they are *katabolic*. Imagine an organism reduced at length by these processes to a formless aggregate of its elemental constituents. Now imagine this formless aggregate of dead material led back step by step till the living organism is set up once more, and you realize the antagonism between organism and environment. For the processes of organisation that preceded death were the precise opposite of all that follow it; they reversed the dissipative tendency of inanimate matter; in a word, they were uphill processes of guidance and direction — were *anabolic*.

The actual relation of a given organism to its environment is usually very complex, the environment in large measure consisting of other organisms. But we shall not go wrong, if, for simplicity's sake, we consider only the physical environment, which is indeed the sole environment of organic life taken as a whole. So doing we see the hopelessness of regarding this environment — which itself is not alive, which antagonises life — as possibly itself the source of life. Neither can we assume a specific vital energy or force, as the old vitalists did; for life has not — so far as we can see — the properties of a definite form of energy. Thus, when life disappears, there is no equivalent amount of other energy appearing in its place, which we might regard as the result of its transformation. We cannot call death a form of energy. Life, in short, seems to consist in the guidance and control of the known forms of energy, molar and molecular. Quite possibly, beside them, there may be unknown forms of energy, but hardly, as commonly understood, such as would explain life itself. For energy — unless there be what might be vaguely called higher forms of it — is directionless, and all physical forces, so to say, katabolic. The progress of knowledge, in fine, discourages all attempts to treat life as a sort of *tertium quid*, mediating between matter and mind. Turning then to the facts of mind, a sound method will lead us first to the daylight of our own conscious experience, not to the glimmering twilight of primitive sentience and instinct. Looking broadly at the facts of mind from this standpoint, we come upon two principles that lead us straight to the teleological factors of organic evolution. One of these is the principle of self-conservation — the wide reach and significance of which Spinoza was one of the first to see;[1] the other is a principle which I ventured many years ago to call the principle of subjective or hedonic selection.[2] These principles furnish natural selection with

[1] Cf. *Spinoza, his Life and Philosophy*, by Sir F. Pollock, pp. 221 ff.

[2] Art. *Psychology*, *Encyclopædia Britannica*, vol. xx, 1886.

the ποῦ στῶ it seems to demand. Without these it is difficult to see what purchase it can have, as I will try to shew presently. But first, a word concerning the principles themselves.

I do not need to weary you with any laboured psychological analysis. It is enough to note that both these principles imply feeling and activity; they imply, too, that the activity is prompted by the feeling. Thus, self-conservation, *i.e.* the conservation of self by self, presupposes the will to live and the pain of dying. It shews itself especially, any unfavourable change in the environment having occurred, in the reactions to this change, which frequently so much exceed the energy of the occasioning stimulus. Apropos of this, organisms are often compared to delicate machinery provided with 'self-regulating' valves, with hair-triggers, and with other devices, for nicely controlling large stores of potential energy or setting it free on slight provocation. No doubt there are many points of analogy between organisms and such ingenious contrivances. But it is forgotten that the said contrivances are themselves invariably the work of mind. Call an organism a machine by all means, if you like; but where is the mind that made it, and I may add, that works it? Descartes, it will be remembered, was content to regard all the lower animals as simply automatic machines, comparable, though superior, to marionette dancers and flute-players such as those made afterwards by Vaucanson, which led Lamettrie to call even man a machine. But Descartes himself stopped short of this, on the ground that the complexity of human manifestations points to what Huxley has since called a conscious, as distinct from a mechanical, automatism. But the inconsequence of Descartes' reasoning has been generally allowed. It was open to him either to refer the greater variety of human life to the great complexity of the human brain, or knowing by direct experience that the human machine was a *conscious* automaton to infer that the simpler machineries of the lower organisms were conscious automata of a simpler type. The explanation of Descartes' inconsistent and illogical doctrine is to be found in the perplexities of the psychophysical problem, with which we shall have next to deal. Led by his fundamental analysis to insist on the complete disparateness of matter and mind, and led, therefore, to reject such hybrid notions as vital force, he saw no way of explaining the interaction of body and mind save by miracle,[1] and naturally was averse to admitting such intervention any further than facts compelled him. His own consciousness, he thought, convinced him that man was a '*mélanges confus*' of

[1] I do not mean that Descartes regarded the union of body and mind in man as continuously maintained by special Divine intervention. His followers were, but he was not, an occasionalist, spite of all Hamilton's contentions to the contrary (edition of Reid, p. 961). This union was for Descartes only 'hyperphysical' in the sense of being a unique fact, a 'negative instance,' as Kuno Fischer aptly calls it. The following extract from a letter of Descartes to Arnauld seems decisive: Que l'esprit qui est incorporel puisse faire mouvoir le corps, il n'y a ni raisonnement, ni comparaison tirée des autre choses qui nous le puisse apprendre, mais néanmoins nous n'en pouvoir douter, puisque des expériences trop certaines et trop évidentes nous le font connaitre tous les jours manifestement. Et il faut bien prendre garde que cela est l'une des choses qui sont connues par elle-mêmes, et que nous obscurcissons toutes les fois que nous voulons les expliquer par d'autres. *Œuvres*, Cousin's edition, x, p. 161.

body and soul; he did not feel forced to say the same of animals or of plants. But if we admit the inconsequence of Descartes' restriction of this concomitance of psychical and physical to man alone among animals; and if we admit, too, the invalidity of treating life as a specific form of energy, — then surely we are bound to assume this concomitance wherever we recognise life. To make my meaning clearer, let me first quote a sentence or two from an essay by a very distinguished botanist, and add one or two comments. The essay is by Professor Strasburger, of Jena, and his subject is *Protoplasm and Sensibility*. Referring to the analogy between organisation and machinery, he remarks: "For the structure of a machine, too, might be called its organisation; and the fact that, when provided with a store of energy, it can be started, by the opening of a valve, to perform work conformable to its structure — this property might be called its sensibility. But the living substance is entirely distinguished from the dead machine by the ability to provide itself with the energy needful for its work; to set itself in motion and keep itself going, to repair itself, within certain limits, the defects that may arise; and, above all, by the fact that it constructs itself. *In short, an organism — in contrast to the dead machine — is a living machine*, one that does not depend on external impulses for its movements, one that regulates its own course and continues going, as long as the environment will allow. Only through the hostility of this or through irreparable misfortune is it brought to a halt."[1] Now, I have said, that wherever we

[1] *Das Protoplasma und die Reizbarkeit*, 1891, pp. 24 f.

see a machine, we ask, Where is the mind that made it, and that works it? In the dead machine this mind is outside and independent; in the living machine, or organism, it is inside, and so far identical. Living machine and conscious automaton are, then, strictly synonymous: whether we say life or whether we say consciousness, we equally imply the development and conservation of self by self through processes working counter to the downhill trend of the physical environment. Looking again at the dead machine, we may ask, What is it made for; what is the work that it is constructed to perform? To crush quartz, roll lead, grind flour, and so on, we are told, as the case may be. But what is the living machine made for? We must answer, be it plant, be it animal, be it man: For itself and for its kind, to live and to multiply. Once more, looking at the dead machine, we find the structure precedes and wholly determines the function; but in the organism, and especially when we take an ascending series of organisms into account, we find it truer to say the function, *i.e.* life, determines the structure.

And so we come to our second principle, that of subjective or hedonic selection.[1] By way simply of illustrating this principle, and deferring meantime all question of its evolutional significance, let me try briefly

[1] There is, I suspect, some considerable resemblance between this principle and that which Professor Baldwin describes as Organic Selection. But as I have not seen all Professor Baldwin's papers and as he does not refer to what I had written ten years before, I cannot speak definitely. If our meaning should turn out to be the same, I should contend for the term Subjective Selection as the more appropriate. Cf. *A New Factor in Evolution*, by J. Mark Baldwin, *The American Naturalist*, July, 1896.

to call up two or three examples. Take the passengers on a coach going through some glen here in Scotland: in one sense the glen is the same for them all, their common environment for the time being. But one, an artist, will single out subjects to sketch; another, an angler, will see likely pools for fish; the third, a geologist, will detect raised beaches, glacial striation, or perched blocks. Turn a miscellaneous lot of birds into a garden; a flycatcher will at once be intent on the gnats, a bullfinch on the pease, a thrush on the worms and snails. Scatter a mixture of seeds evenly over a diversified piece of country; heath and cistus will spring up in the dry, flags and rushes in the marshy, ground; violets and ferns in the shady hollows, gorse and broom on the hilltops. I am aware, of course, that thrushes and flycatchers, flags and heather, are products in large measure of natural selection, that is of what we have agreed to call a non-teleological factor. But I do not think this will be found to militate against these examples for my purpose. The complete unravelling of the two sets of factors, teleological and non-teleological, so as clearly to exhibit their respective shares in any given form is probably an impossible task. My concern is only to show that the two sets of factors are there, and that the teleological are indispensable. It will suffice then to observe that by the principle of subjective selection special environments are singled out by different individuals from the general environment common to all, and that so far there is not necessarily any competition. Two artists or two anglers may be in each other's way, but an artist and an angler will hardly incommode each other. A garden would still interest a flycatcher if there were neither pease nor cherries in it, provided the insects remained; whereas the bullfinch would at once forsake it. Natural selection as distinct from subjective selection comes into play only when two anglers contend for the same fish, two artists compete for the same prizes, when the early bird gets the worm that the later one must go without.

Let us next put this principle into shape and then we may consider its evolutional significance. Psychologically regarded, movements are determined by feeling: indifferent sensations, therefore, that occasion no feeling, lead to no movement in response; while the same presentation, if it occasion opposite feelings in two different individuals, will be followed by contrary movements. As I have put it elsewhere: "The twilight that sends the hen to roost sets the fox to prowl, and the lion's roar which gathers the jackals scatters the sheep. Such diversity in the movements, although the sensory presentations are similar, is due," then, to the fact "that, out of all the manifold changes of sensory presentation which a given individual experiences, only a few are the occasion of such decided feeling as to become objects of possible appetite or aversion."[1] So we may formulate our principle; which granted, certain important consequences follow deductively when we connect it with well-known psychological laws. Specialisation means also concentration; the more restricted the lines of reaction, the more perfect these reactions become. The "Jack of all trades is master of none." Thus sub-

[1] *Encyclopædia Britannica*, article *Psychology*, p. 42.

jective selection will determine definite variations as distinct from fortuitous ones, definite in the sense of bringing the individual into closer *rapport* with that portion of the general environment which it is selecting.

And now let us reflect how much these principles mean. Natural selection, it is allowed, is metaphorical. The common environment is not an agent, and selects as little as it conserves. Its tendency, if we consider it alone, is not to produce variations any more than to produce life; on the contrary, its tendency is towards uniformity and quiescence, as we may see in the dust and ashes to which in the end it reduces all. But in subjective selection there is nothing metaphorical; the agent here — so at least we must say as psychologists — is real, the source and type of all our conceptions of activity. I do not forget the psychophysical inquiry still pending; but that in any case has to accept psychological facts, being merely a theory about them. The agent then is real, not an abstraction; the selection likewise is real, not metaphorical. The individual positively selects what is pleasant, that is what conserves, for appetition; and negatively selects what is painful, and so detrimental, for aversion. To the remainder it is indifferent. By such selection is constituted its proper and *specific* environment. The origin of this kind of species, species of environments, at any rate seems due to a psychical, not to a physical, selection. Moreover, there is so far no struggle for existence, where "all subsists by elemental strife": rather here, as the poet has said, "All nature's difference keeps all nature's peace."[1]

[1] Pope, *Essay on Man*, i, 169; iv, 56.

So far we may get by connecting our principles with the well-known psychological law, that concentration and practice perfect functions, and the corresponding physiological law, that function perfects structure. But there is another psychological generalisation with which I think we may connect them, and which imparts to them still further teleological significance. We have found Darwin exclaiming against "Lamarckian nonsense of a tendency to progression." But if nonsense, it is nonsense of which many great thinkers have been guilty. We find it, of course, in "the wisest of wise Greeks, the Stagirite," and in our day — spite of Darwin's disclaimer — it is still avowed by such leading biologists as Nägeli, Kölliker, and Virchow. No doubt Aristotle's conception of an internal perfecting principle was vague and lent itself to mystical interpretations. But I believe the progress of psychology will enable us some day to give it greater definiteness and a more assured foundation. Meanwhile time forbids any attempt to work further at this point now. But I will venture to quote a few sentences of my own published ten years ago, that may suffice to indicate what I mean: "How in the evolution of the animal kingdom do we suppose this advance from lower to higher forms of life to have been made? The tendency at any one moment is simply towards more life, simply growth; but this process of self-preservation imperceptibly but steadily modifies the self that is preserved. The creature is bent only on filling its skin; but in doing this as pleasantly as may be, it gets a better skin to fill, and accordingly seeks to fill it differently. Though cabbage and honey are what they were before, they have changed relatively to the caterpillar now it has become a butterfly. So, while we are all along preferring a more pleasurable state of consciousness before a less, the content of our consciousness is continually changing; the greater pleasure still outweighs the less, but the pleasures to be weighed are either wholly different or at least are the same for us no more. What we require, then, is . . . that to advance to the level of life on which pleasure is derived from higher objects shall on the whole be more pleasurable or less painful than to remain behind." [1] Now this condition seems provided, without any need for a clear prevision of ends or any feeling after improvement or perfection as such, simply by the waning of familiar pleasures and by the zest of novelty. In the midst of plenty it is usual to become more dainty and to make efforts to secure better fare, even though the old can be had without them. Exceptionally no doubt such circumstances lead to an opposite result, as we see in the degradation of most parasitic forms. But the principle of self-conservation seems sufficient to render this result exceptional.

Thus—even if there were no natural selection of variations fortuitously occurring, and even if there were no struggle for subsistence, still—the will to live, the spontaneous restriction of each individual to so much of the common environment as evokes reaction by its hedonic effects (with the increasing adaptation and adjustment that will thus ensue), and, finally, the pursuit of betterment to which satiety urges and novelty prompts, —these conditions, really implying no more than the most rudimentary facts of mind, will account for defi-

[1] *Encyclopædia Britannica*, article *Psychology*, p. 72.

nite variations to an apparently unlimited extent. What is more, the variations so produced, even if there were no others, would furnish natural selection with an ample basis as soon as struggle for existence began. They would also remove or minimise one of the most formidable difficulties in the way of natural selection working alone — a difficulty which Mr. Herbert Spencer has had the credit of pointing out. It is easy to imagine a single variation which is at once useful, occurring fortuitously; and it is plain that natural selection will secure its survival. But when, as Darwin allows to be generally the case,[1] utility depends on the coördination of a number of variations separately useless, then the chances against the simultaneous occurrence of these in due correlation increase at an alarming rate as the number of independent variants increases. Proportionally large drafts on time thus become requisite before such complex utilities can arise by lucky accident. We might say, I think, that not only are geologists accused of asking more time than according to the astronomer's facts the physical history of the earth will afford them, but that the demands of ultra-Darwinians like Weismann may expose them to a like charge on the part of geologists. Weismann long ago expressed the hope that at no distant date he would be able to consider this objection — I mean the difficulty of coördinations; but, so far as I am aware, he has not yet made good his promise.

The mention of Weismann's name reminds me that many of you will be thinking of his famous doctrines of heredity and germ-plasm. If those doctrines are true, it

[1] Cf. *Origin of Species*, sixth edition, pp. 178 *fin.*

will be said, acquired characters cannot be inherited, and the Lamarckian and other like teleological factors become so far impossible. As to the truth of Weismann's properly biological doctrines I have no right to express an opinion, but there are some characteristics of his method on which I may remark. *First*, 'acquired' and 'congenital' do not seem to be terms whose meaning is independently fixed. If a character turns out to be inherited, Weismann thereupon feels entitled to call it congenital, even though he had previously in common with the rest of the world regarded it as acquired. Speech, for example, is an instance which he himself selected as an acquired capability, urging accordingly that the human infant ought to begin by talking. When it was pointed out that it does begin by "babbling articulate syllables," the Weismannians urged, if Romanes may be trusted, that after all, "seeing of how much importance this faculty must always have been to the human species, it may very well have been a faculty which early fell under the sway of natural selection, and so it may have become congenital."[1] *Secondly*, it must be frankly admitted that in many instances in which acquired characters have been said to be inherited or might be expected to be inherited, the Weismannians have shown that nevertheless there is no such inheritance. But induction by simple enumeration is not sound logic. What the theory requires and assumes is the absolute non-inheritance of any acquired characters—a negative obviously difficult to establish. On the other hand, to overthrow the theory, it suffices if its opponents can shew that in any particular instances

[1] *Darwin and after Darwin*, vol. ii, p. 336.

acquired characters are inherited. Several such instances have been adduced, and Weismann is at this minute devoting all his ingenuity to explaining these instances away. *Lastly*, in so doing he is driven not only to modify his theory, but to render it more and more cumbrous, complicated, and artificial. The more the body-plasm is eliminated as a medium of heredity, the more wonderful and miraculous the germ-plasm becomes. 'Ids,' 'idants,' 'biophores,' 'determinants,' have an obviously teleological ring and yet are meant to make the teleological superfluous. They remind one of Mr. Spencer's speculations concerning organic evolution referred to in the last lecture; indeed, Weismann himself admits the resemblance. Yet, spite of the proverb that people in glass houses should not throw stones, we have the odd spectacle of Mr. Spencer vigorously bombarding Weismann's bulwarks, quite unconscious of the fact that he is thereby seriously damaging his own.

We seem warranted, then, in concluding, with Darwin himself, and Weismann notwithstanding, that natural selection without teleological factors is *not* adequate to account for biological evolution; and further, that such teleological factors imply not a nondescript force called vital, but a psychical something endowed with feeling and will. Finally, recalling our survey of evolution in the wider sense, we have seen that, unless the cosmos itself is to be regarded as a finite and fortuitous variation persisting in an illimitable chaos, we must refer its orderliness and meaning to an indwelling, informing Life and Mind. But the problem of the relation of Mind to Mechanism still remains.

For EU product safety concerns, contact us at Calle de José Abascal, 56–1°,
28003 Madrid, Spain or eugpsr@cambridge.org.

www.ingramcontent.com/pod-product-compliance
Ingram Content Group UK Ltd.
Pitfield, Milton Keynes, MK11 3LW, UK
UKHW010350140625
459647UK00010B/956